T0303915

Human Factors in System Design, Development, and Testing

Human Factors and Ergonomics

Gavriel Salvendy, Series Editor

Human Factors in System Design, Development, and Testing

David Meister
Consultant

Thomas P. Enderwick
Space and Naval Warfare System Center

CRC Press
Taylor & Francis Group
Boca Raton London New York

CRC Press is an imprint of the
Taylor & Francis Group, an **informa** business

CRC Press
Taylor & Francis Group
6000 Broken Sound Parkway NW, Suite 300
Boca Raton, FL 33487-2742

This book is dedicated to the memory of
Shirley Davis Meister
and to
the often unappreciated HF specialist.

Contents

Series Foreword

With the rapid introduction of highly sophisticated computers, telecommunication, and service and manufacturing systems, a major shift has occurred in the way people use and work with technology. The objective of this book series on Human Factors and Ergonomics is to provide researchers and practitioners alike with a platform where important issues related to these changes can be discussed, and methods and recommendations can be presented for ensuring that emerging technologies provide increased productivity, quality, satisfaction, safety, and health in the context of the information society.

This book eloquently and convincingly presents the notion that the chief function of the Human Factors and Ergonomics profession is to attempt to modify data technology through the application of behavioral science principles to make it more humane. By so doing, not only will the users of the technology be more satisfied, but also their performance will improve with respect to reduced error rate and accident rate, and performance output will increase. These factors will provide a valuable and convincing argument for the consumers to buy these ergonomically well-designed products and processes because they will be better and frequently cheaper than those not exposed to system design.

In the eight chapters of the book, the user, processes, method, and practice of system development are presented, and the impact of cross-cultural factors on system development is discussed. A special emphasis is placed on software design, but the critical thrust of the book is the provision of a framework for understanding, controlling, and designing the symbiosis of human and artifact. The book provides a well-elaborated, theory-based, practically oriented, thorough process for achieving this objective.

–Gavriel Salvendy, Series Editor

Preface

Human factors (HF), as a discipline, began in 1941 as an application of the principles of experimental psychology to the design of war equipment. It has expanded over the years into a distinctive discipline with many subareas (Meister, 1999). One of these areas — one crucial to HF — is the contribution of HF to design.

Many professionals, however, still see HF in design as simply an application. Our point of view is somewhat different. Unless the design relationship assumes a central role in HF, the latter is simply a variation of psychology. HF is actually much more than that.

HF in design is, of course, an application of behavioral principles and data, but it has certain elements that make it, as a discipline, unique. If technology is important to our present civilization (and few would deny this) and if design is the means to implement that technology, then HF as a discipline has the responsibility to study how human performance is affected by technology and, equally important, how behavioral principles can and do influence that technology. The latter is the subject matter of this book. Let us state it boldly: The study of HF in design is a basic function of the discipline.

Despite its physical context, design is essentially a behavioral phenomenon. It is a creative activity like composing music or painting, with the important addition that design is fundamentally cognitive and rationalistic. All of this makes design a significant topic for HF study.

For the HF analyst, the major question is how the human uses behavioral principles to influence the form of a physical equipment. This is the problem of behavioral/physical transformation, which is akin to philosophy's mind–body problem.

There is also a certain ethical component to the application of HF to design. The effect of technology can often be stress-inducing, even crushing, to the technology user. We say that HF has a responsibility to do what it can to ameliorate the negative effects of technology — to make it more humane. It cannot do this without seriously studying the relationship between the human and technology, a relationship manifested mostly in design.

If the reader finds some of the following material difficult to follow, it is because the questions raised are complex. There are no simple answers, and the number of questions discussed is more than the number of answers provided. This, however, is a characteristic of science.

A few words about what this book is not: It is not a "how-to" guide; it is not an introductory text for the first-year undergraduate; it does not provide step-by-step procedures that, if followed to the letter, will solve all design problems. The relationship of HF to system design is too complex to be encapsulated in a series of aphorisms like "keep it simple, stupid." What we are interested in — and we hope the reader is too — are the thought processes that comprise the foundation of what the HF professional working in system development does.

— David Meister
— Thomas P. Enderwick

List of Acronyms

AFB	—	air force base
AI	—	artificial intelligence
AR	—	analogical reasoning
ATM	—	automated teller machine
CAD/CAM	—	computer-assisted design/computer-assisted manufacturing
CE	—	cognitive engineering
CIA	—	Central Intelligence Agency
CIC	—	Combat Information Center
CTA	—	cognitive task analysis
D/A	—	decision/action (diagrams)
DM	—	decision making
DOD	—	Department of Defense
DSS	—	decision support system
EID	—	ecological interface design
FA	—	function analysis
GOMS	—	goals-operators-methods-selection (model)
GUI	—	graphic user interface
HCI	—	human computer interface
HF	—	human factors
HFE	—	human factors ergonomics
HFES	—	Human Factors and Ergonomics Society
HMI	—	human–machine interface
ICBM	—	intercontinental ballistic missile

mph	—	miles per hour
ms	—	milliseconds
NPP	—	nuclear power plant
OE	—	operational environment
OPTEVFOR	—	Operational Test and Evaluation Force (Navy)
OSD	—	operational sequence diagrams
OST	—	operational system test
PC	—	personal computer
RP	—	rapid prototyping
SEP	—	system, equipment, product
SME	—	subject matter expert
S–O–R	—	Stimulus–organism–response
SPAWAR	—	Space and Naval Warfare Systems Center
SRK	—	skills, rules, knowledge (theory)
TA	—	task analysis
TD	—	task description
T&E	—	test and evaluation
UT	—	usability testing
VDU	—	visual display unit
WW	—	World War

1

An Overview of System Development

This book logically begins with a list of questions regarding human factors (HF) in design, development, and testing. Some of these questions cannot be answered at all, many of them can only be answered partially, but all of them require analysis.

The questions, which are discussed in the next section in greater detail, have merit not only because they can serve to organize the topics of the remainder of the book, but because answers to these questions help us improve the effectiveness of the HF specialist in system development.

A few definitions first. From now on, when we refer to the *specialist*, we mean the HF personnel working in design, development, and testing. We differentiate the specialist from the engineer (who may be one of several types: electrical, hydraulic, logistical, etc.) and designate him or her as the *designer*. Unless specified otherwise, we assume that the specialist is not the primary designer of a system (how much he or she contributes to the human–machine interface design undoubtedly varies), although he or she is probably a member of the design team. It is necessary to keep these two — specialist and designer — separate because the all-inclusive terms *design* and *development* tend to confuse their individual contributions. Also, rather than repeating the terms *design*, *development*, and *testing* throughout the book, we use the term *design* to refer to all these stages.

The questions for which we would like answers are:

1. What are the critical elements in system development, and who are the people involved?

2. How do designers design? What theories and models have been formulated to describe design and development?
3. What informational resources (e.g., design guidance, facts, data) do designers have, and how is the design guidance created? What information do designers want and need?
4. What is and should be the role of the specialist in system development? What information does the specialist require?
5. What are the criteria and attributes that distinguish a good design from one less good?
6. What should HF research do for system development, and what should be the characteristics of that research?
7. What are the significant differences between hardware and software design and between commercial and noncommercial design?
8. What should the user's role be in system development? What information do we wish to secure from users, and how can it be elicited? How does user-centered design differ from any other kind of design?
9. What are the characteristics of the development process, and what theories have been constructed about it? What is the role of constructs like indeterminism and complexity?
10. What methods of analyzing design requirements and testing systems at various levels of complexity exist, and what information does each provide?
11. How does the organizational context (including the type of design problems to be solved) affect development?
12. How extensive are individual differences among designers and design practices, and what causes these differences?

It should be obvious that these questions are *meta-level* questions, and each subsumes a host of more specific ones.

This chapter is concerned with the nature of system development. For the individual specialist and designer, this is only a context for what he or she does at the design or drafting table. Although the overall developmental process affects what individual personnel do in the process, the effect may appear indirect or insubstantial. Nevertheless, because development as a whole does have an effect (however indirect) on individual design efficiency, it is necessary to examine what theorists have thought about it.

The nature of the equipment or system being designed (we use the general term *system* to represent all levels of equipment size and complexity) and the design problem presented also have an important effect. At a certain level of complexity, systems develop certain attributes they did not have at lower levels of complexity (e.g., automation, transparency, autonomy, etc.), which must also be considered.

DEVELOPMENTAL ELEMENTS

The following are not in order of importance or in any logical sequence. Some elements may be more important than others, but this appears only with further consideration. Having briefly noted what the system development elements are, we examine each in greater detail later. Each of these elements implies a question: what, how, and why?

The Nature of the Overall System Development Process

We differentiate between the design process and system development; the latter subsumes the former because it is more comprehensive. For example, development includes continuing testing and evaluation of design outputs like software scenarios and hardware drawings. This results in iterative feedback loops that modify the original design. Design takes place in a context that is development and testing. Consequently, one must try to understand how the development context proceeds and affects the design of the individual equipment and system. One wishes to determine the structural characteristics of the design process because this understanding provides, to some extent, the ability to control them. We wish to know whether these processes are highly rational or influenced by irrational factors: Does design proceed from top to bottom, bottom to top, or both? What are the stages of development and characteristics of each? Is the design new, an updated version of an earlier system, or a redesign, and what effect does this have?

The Nature of the Equipment/System to Be Designed

We draw distinctions among systems, equipments, and products. An *equipment* is the individual work station; the system is composed of a number of work stations, all functioning together for a concerted purpose or mission. A product is something developed to perform a function; a product does not have a purpose or mission. These distinctions are hardly rigid; they meld into each other. For example, an automobile is certainly a product; however, although it is not a work station (or is it?), it is an equipment and may even be considered a system because it contains subequipments or modules. The same piece of hardware or software may be considered in one or other category depending on whether the individual encountering the device is an operator or technician. In general, in this book, we are concerned more with equipments and systems, but whether that device is fish or fowl depends on the user's point of view. When it is difficult to even establish basic categories, the complexity of the problem we examine is revealed.

Whatever one wishes to call the device, however, certain factors determine what we call these machines (e.g., whether the machine performs a function or has a purpose given it by its operator—only a system has a purpose; an equipment has only a function). Another factor is whether a device is an output of another device, in which case we call it a *product*. For example, a can has a function (to enclose an object), but cannot be considered an equipment or a system, although the machine that makes the can is certainly considered an equipment.

Because we are interested in the process that leads to an output, but not necessarily the output itself, we confine ourselves to equipments and systems. Yet products are technologically derived, and more people are involved with products than are involved with equipments and systems. Yet one must confine oneself to what is feasible. Although it is difficult to make the distinction because the terms *equipment* and *system* are often used interchangeably, nevertheless the system has attributes other than those of the individual equipment and produces effects that are not those of the equipment. These system attributes include a greater degree of complexity (because of the multiplicity of equipments) and a mission.

Certain parameters associated with the nature of the equipment/system strongly influence the details of the design process. Among these are the degree of complexity and automation involved; the degree of transparency and autonomy; whether the system is to be military, industrial, or commercial; whether the system is to be largely hardware or software or both; and the specific type of system. Later we examine other factors associated with the equipment/system type that influence the design.

The Nature of System Design

This is a matter of high-level conceptualization or abstraction because no one individual can really define the design process; it is so big that only by making it into abstraction can one tie the process down. Hence, conceptualizations of the design process are like philosophical propositions—like the scholastic philosophers in the Middle Ages worrying about how many angels can stand on the head of a pin. Because the design process is under no one's control (even that of the engineering manager), it contains everything, and consequently it is essentially our perception of phenomena.

The Characteristics of Design Personnel

Design (and here we include the specialist's activities as well because he or she is intimately involved in the design team) is essentially idiosyncratic because it is informal (being a problem to be solved). Theorists can propose formal methods to be followed by the design process, but whether these are implemented depends on the individual designer, hence his or her skills, knowledge, and attitudes (and even such rules as exist). Other factors include the information available to the designer, customary practices that depend largely on personal work experience, and the constraints under which designers work.

We include engineering management as another personnel element in design. However, in contradistinction to the individual specialist/designer, management is concerned only rarely with the design of the individual work station or equipment, but mostly with the system development program as a whole (its scope and protocol).

Information Resources

Some theorists (e.g., Rouse & Boff, 1987) view the entire development process as being organized around information transmission, reception, and utilization. These resources include: (a) military and commercial specifications; (b) design guidelines or principles of behavioral design; (c) prior system documentation; (d) the way in which specialists and designers seek out and utilize information; (e) the impact, if any, of behavioral research published in the literature; (f) the questions that designers ask and for which they seek information; and (g) how specialists conceptualize their discipline. One must also consider the utility of information, its validity and reliability, and the use of subject matter experts (SMEs) as information resources.

Formal and Informal Methods

Although there are many behavioral methods for design analysis, development, and testing in the literature, it is unlikely that all of them are used by all specialists because method use interacts with type of design problem, money and time availability, and so on. Therefore, it is necessary to examine the characteristics of these methods, their advantages and disadvantages, as well as the extent to which they are utilized by specialists (it is assumed that designers do not make much use of behavioral methods). We must also be concerned about the effect of computerization on design and testing.

Design Inputs and Outputs

Design cannot be understood without examining inputs and outputs. Design is, of course, fundamentally a process, but it receives inputs and produces outputs that must be examined to understand the overall process. Sample inputs to the design process are design specifications produced by a customer who apparently knows what he or she wants; sample outputs from that process are scenarios developed as part of requirements analyses. There are, of course, engineering drawings in hardware development and storyboards in software development. Design inputs are likely to be more constraining to the designer than design outputs because the designer has more control over the latter than the former. Rules and procedures operate on outputs more than they do on inputs. If design is in part a process of reacting to stimuli (inputs), we must pay more attention to the former than the latter.

The User

The operator (customer) of the design product has become a much more visible design element in the past 10 years. Many years ago, the user's role (as a company, governmental agency, general public customer) was to present to engineering a challenge that the designer would solve, largely independent of inputs from the user during the design process. If design was successful, the user was pleased; if he or she was not pleased, the designer eventually lost the particular user.

A user is always one who employs an object, but actual users are by no means monolithic. The user may be a consumer or purchaser of a commercial product; a user may also be an operator of an equipment that he or she often does not own. A user is someone (or some company) who contracts for another to design/construct an object to his or her desires. One may use both physical objects and symbols such as information. A user may also be someone who does nothing except take advantage of the outputs of a device (e.g., being a passenger in an aircraft). A user can provide inputs to design about him or herself, others, and what they do; a user may be part of a developmental process or may be completely remote from that process.

The role of the user/customer in the development process is important if only because that development is initiated by the attempt to satisfy customer/user desires. At this point, it is necessary to make another distinction. There may in fact be (initially at least) no immediate customer for an equipment or system. Each major engineering company has a new development group whose personnel spend their time trying to develop a new machine or process that as of yet has no customer. The company creating this novelty must find a customer for it or someone to sell it to after the design.

Much of the computer industry is of this nature: Engineers develop a new mouse trap and attempt to induce consumers to buy it. Examples are high-resolution TV or computer search engines that scan millions of items in a second. No customer ever showed up for these design improvements until after the design had been largely completed. This has always been true of technology. The advantages of these improved designs may be somewhat dubious (considering the nonsense presented on TV, why should one require a high-resolution device to watch it unless one is a sports fan?). However, the point is that users do not come forth to direct the design process unless one considers the engineer who created the new device as a user. Indeed, sometimes the new device has difficulty finding a user. For example, the Gatling machine gun developed during the American Civil War did not initially find buyers, and many improvements in guns following its development were stoutly resisted by various armies.

The notion that the designer can be viewed as a user makes it critical that we extend to the design situation the same scrutiny and attention that we extend to the consumer as user. In fact, it is arguable that the designer as user requires more attention than the consumer as user. User reactions to prototype designs can also supply valuable feedback. Increasingly over the years, efforts have been made to include customers/users in the development process because without their involvement it is possible for design to go seriously astray.

Although recent design-relevant literature in HF focuses on the user, we know very little about him or her. Perhaps advertising companies, which are constantly sending out questionnaires to users, know more. Usability testing and rapid prototyping center around users, but the role of the latter is very much that of the response object in a test program. One tries to find as much as possible about what the user wants before design, but design cannot respond uncritically to user desires even if the user can articulate these. We need to know much more about the user; but in learning about users, we must be more skeptical of their value vis-à-vis design.

THE REMAINDER OF THE BOOK

The remainder of the book discusses each of the seven design elements. Chapter 1 continues the discussion of the nature of system development as well as the nature of the equipment/system. Chapter 2 describes the design process. Chapter 3 highlights design methods. Chapter 4 reports on design practices. Chapter 5 provides information resources (e.g., available HF research, design guidelines, and standards). Chapter 6 deals with software design, Chapter 7 with the user. The book concludes (Chapter 8) with a description of a behavioral theory of the design process.

CONCEPTS OF THE SYSTEM DEVELOPMENT PROCESS

Traditional Concepts

A number of theorists have attempted to characterize (describe/analyze) the total developmental process. We hypothesize that this is an effort to exercise more control over the process, which when one encounters it physically for the first time appears to be somewhat disorganized. The neophyte specialist may view the process as frenetic, confusing, and somewhat irrational.

For a new complex system, design takes place on multiple levels. Presumably it all interacts, but the surface appearance is one of confusion, although it may be organized confusion. The attempt to understand this confusion impels system development theorists to develop abstract principles describing that development — the assumption being that if one can understand it, one can presumably control it more effectively.

System development is a process that consists of design, testing, administration, and management. It can be conceptualized on several levels. At the primary level, an individual engineer is absorbed in a single problem (e.g., to develop an electrical module that can sense certain flight parameters and that, if these deviate from preset values, will energize a destruct mechanism). At a higher level, an entire electrical group is concerned with creating a system that will guide a missile 4,000 miles to a specified destination. At a still higher management level, a bureaucracy exists to ensure that the parts of the engineering system interact and that development goals are satisfied. All of this activity also involves secretaries, documentation specialists, mechanics, carpenters, and so on. Obviously design and that part of it that concerns the HF specialist and the operator–equipment relationship is only one part (possibly only a small part) of the management of a complex, highly dynamic process.

This detail should help the reader understand that if one could stand apart from that process and view it from various angles, the impressions one might receive might be different from each angle. The observer might receive several impressions: It is a frenetic, somewhat uncontrolled and uncoordinated activity; it is a controlled intellectual activity guided by a system requirement; it is an information-seeking and utilization process; it is highly creative and therefore a somewhat idiosyncratic behavioral phenomenon. Depending on where one stands, all of these impressions are correct, but the totality of these impressions — system development viewed as an abstraction — may be different from any one of them. That is because development is, at one level, a set of very concrete operations and, on another, a theoretical abstraction.

Leifer (1987) suggested that *playfulness* is an important attribute of design activity. Design is viewed as a problem, and problem solution requires the creativity of the designer. There is an overriding management structure to the process, but for the individual designer the problem can be viewed in various ways and perspectives: Information can be sought for and utilized in various ways; hypotheses can be created and accepted or rejected. All of this gives the appearance of lack of structure, but this may be the product of our lack of understanding rather than, as Leifer suggested, an inherent randomness.

Along with Rouse and Boff (1987), we view design and development as a problem to be solved on various levels. There are, of course, specifications (in more or less detail) of what is wanted, but how to achieve these goals is often unclear. There are three types of design: new design, update, and redesign. New (initial) design has no prior system that can serve as a model (although systems of the same general type often exist), and so the problem is most difficult. In update and redesign, the first to supply additional capabilities and the second to remedy known deficiencies, there is a predecessor system whose outlines can serve as a template for the changes to be made. As we see later, there are prior system data, test results, and so on that form the basis on which to build; in new design, although there are precedents from a general class of system, the specific design at issue has far less to guide it. Of course, as technology — primarily computerized information — seemingly spins out of our control, the problem is intensified.

In general, despite whatever prior information one has, the traditional design process fits the following paradigm:

1. formulation (analysis) of the problem;
2. development or synthesis of possible solutions, which can be thought of as alternative hypotheses;
3. analysis and evaluation of these alternatives;
4. selection of one alternative that is considered the most desirable problem solution;
5. implementation of the selected design option in detailed design; and
6. evaluation and testing of the selected design alternative.

This last step (evaluation and testing) is an essential part of a logical design process, but management decree may be substituted for empirical testing because of cost or potential risk factors (i.e., management may decide that one of a number of design solutions proposed by engineering is probably the correct one and let it go at that). This is, of course, a high-risk course of action, but management is often attracted to short-term and short-sighted solutions.

The paradigm seems logical but is weakened because at each stage the design team (unless the design problem is very simple, there is usually a team) lacks all the information it needs, may interpret that information inadequately, and allows experiential biases that play an undue role in the selection and evaluation of the problem solution.

The paradigm is greatly affected by many variables. We already mentioned the three types of design: initial, update, and redesign. The type of design influences the efficiency of the paradigm because it influences the amount of information available to the designer. Clues are most visible in the case of the redesign because what is to be redesigned has been suggested by prior experience or testing, which indicates what must be redesigned. In the case of the update, one is not dealing with complete novelty; one begins with an existent system. Thus, the definition of the problem has significant consequences. The new design, as compared with the others, is relatively ill defined; because of this, the danger of producing an inadequate design is greater.

Earlier we distinguished between design of a single equipment and design of a suite of interactive equipments serving a common purpose; we call this a *subsystem* or *system*. Obviously the latter is more difficult than the former, not only because more is involved, but also because the system may have certain attributes not necessarily found in a single equipment.

Design is considered to have three phases: analysis of the design problem, design as an attempt to solve the problem, and evaluation and testing to ensure adequacy of that solution. Along with administration and management, the three phases are called *development*.

Meister (1987) produced a conceptual model of the design process (Table 1.1). Unfortunately, there is no accepted theory of design. However, there is some design literature. Leifer (1987) suggested that this can be categorized into three types. The earlier literature describes the psychology and behavior of individual designers (Amabile, 1983; McKim, 1980; Osborn, 1963). It includes a taxonomy of design methods and strategies (Jones, 1980; VanGundy, 1981). A second body of literature deals with creativity (Adams, 1984). The most recent source of thinking about design has been strongly influenced by computer science and artificial intelligence (AI) concepts. These theorists are interested in formal models of design (Mostow, 1985), knowledge as used in design (Bobrow, 1985), and human– computer interaction (Card, Moran, & Newell, 1983). All of these concepts are considered the traditional way of looking at design. Later we examine newer concepts.

TABLE 1.1
A Conceptual Model of the Design Process

Questions Asked	Design Mechanism
1. *Analysis of the design problem*	
What are the system, subsystem, equipment, and module supposed to do?	Mission analysis
What are the functions and parameters of the system, subsystem, equipment, and module as related to the mission?	Engineering knowledge; deductive logic
What already established system elements are fixed? What is their effect on other system elements?	Examination of system requirements
What information about the system is known? What is unknown?	Knowledge of the state of the art and of the test literature; gathering of additional information
Do any aspects of the problem resemble those the designer previously encountered? Are any previous design solutions applicable to the present problem?	Designer experience
2. *Generation of alternative solutions*	
What are all the possible solutions that can be applied to this problem?	Deductive logic: engineering knowledge; designer experience
3. *Analysis of alternative solutions*	
What criteria apply to this problem?	Engineering knowledge; designer experience
What constraints impact the alternatives?	System requirements; information from management
What alternative parameters can be traded off?	Engineering knowledge
What are the similarities/differences among alternatives?	Engineering knowledge
4. *Selection of preferred solution*	
What are the advantages/disadvantages of each alternative?	Engineering knowledge; design and test data
Which alternative is the best choice?	Paired comparison of alternatives

Before discussing these concepts, it is necessary to distinguish between what the individual designer or design team does (in meetings, discussions, developing drawings, etc.) and what the overall development project does (integrates a number of individual work station designs into a suite that represents the total system). Engineering management obviously has overall control over the individual design efforts, but the control is limited because it is hierarchical, and the chief engineer cannot know what designer X does for his or her part of the total effort.

That management control may be lacking is exemplified by an experience the senior author had when he worked on the Atlas ICBM. There came a time when it became necessary to meld the individual launch systems into an overall procedure that would permit the launch. Management gathered representatives of the individual engineering subsystems (e.g., electrical, hydraulic, logistics) into a single room and attempted to review a preliminary launch plan, with each engineering specialty invited to describe its part of the procedure. After the first four steps in the procedure, it became apparent that no engineering department had been communicating with any other, so there were many unknowns (e.g., is there a switch to accomplish X function?). We were dismissed from the meeting and a number of special liaison engineers were appointed to perform the integration effort that management had originally thought was inherent in the development process.

In an attempt to achieve more control (at least on the theoretical level), a number of concepts to describe system development have been created. These can be divided into two types, although the distinctions are not necessarily rigid: descriptive and analytic.

Descriptive Concepts

The most common descriptive concept is what is termed *top–down* design. This presumes that design is sequential and inherently logical, that it begins with some human function that must be performed to solve the design problem, and that a number of design options are available (or are created by the designer). From these options, one is selected or, in situations in which there is no human function indicated, the designer may decide whether a human or machine instrument will perform the function, and from this certain logical steps proceed.

Top–down design is considered to be inherently rational, emphasizing what some call if–then thinking: If a design option is selected, then certain consequences flow from that selection and there are more detailed possibilities, from which again a choice must be made.

It is obvious to the observer that design is iterative. Either tests at a more detailed level produce feedback that requires the designer to modify original ideas or further analysis at more detailed design levels suggests that it would be wise to revert to an earlier level of design and modify it.

Development involves a proliferating body of choices as one proceeds to more molecular levels. At the same time, earlier, higher level decisions constrain the more detailed decisions made at more molecular levels.

The concept that one can dichotomize the development concept into absolutes is farcical. Development involves both top–bottom and bottom–top decisions. How much of each may be determined by the nature of the design process: New design may require more top–bottom thinking, whereas redesign may require more bottom–top thinking; but both types are involved in all design.

Development as Problem Solving

Any analysis of the design process indicates that all design is an effort to solve a problem that is presented by a customer or user requirement. A system must be developed to produce certain effects. However, such a system does not presently exist—ergo, a problem must be solved.

Conceptualized as a problem-solving process, development requires the designer to analyze the problem presented in terms of its implications for both human and machine functions. Problem solving involves a number of subordinate elements: analysis of the stimulus situation (the customer specification, the assumed user needs), development of hypotheses about the mechanisms that most effectively implement required functions; application of experience (usually at the individual designer level) to the problem solution (i.e., what has worked in the past for a similar problem—this may involve going for the first solution that works); and development and analysis of a number of design options, as well as testing of these conceptually and by developing prototypes and physically testing these. One design option is selected and subjected to more intensive conceptual and physical testing; this process goes through a series of iterations, and the completed system is exposed to interim and final operational or usability testing.

In this scenario, design does not differ significantly from the solution of any other problem, mathematical, physical, chess, and so on. Those who observe the actual design process would like to add another element to the problem situation: intuition or creativity. Few observers of actual design would claim that design is a completely rational process. Indeed, when specialists are asked about this (see Chapter 4), the preponderance of responses suggests that intuition does enter into design; the only question is to what extent. It is highly likely that design/development is like most real-world processes—multidetermined—and there is a shifting trade-off of factors and their effects.

Analytical Concepts

These concepts include those that emphasize information processes (Rouse & Boff, 1987) and communication of information. It is also possible to think of the development process in economic terms—as a series of trade-offs and negotiating compromises. Certainly, each design factor may have advantages in functionality, together with associated costs. Because each system design process usually contains a number of possible options, the task of selecting one option or at least winnowing down the number of options to a precious few requires trade-offs and negotiations.

This is especially true because the design team contains separate interests (stakeholders), each of which must be considered.

Nontraditional Concepts

What has been discussed so far is a design process that describes how systems were developed in 1950. Does it apply to the computerized systems of the 1990s? Does it represent the newer thinking, which is centered around the user and the user's concerns?

A number of new, nontraditional concepts of design have been advanced. The new thinking goes by a number of names: *social-centered design* (Stanney, Maxey, & Salvendy, 1997), macroergonomics (Hendrick, 1997), scenario-based design (Carroll, 1995).

Stanney et al. (1997) contrasted what they call *Taylorist system-centered design* with *human-centered design*. The former presumably considers humans as only resources to create more productive systems (thereby enhancing efficiency and profits). Human-centered design considers the capabilities and limitations of humans and provides opportunities for psychological growth of the individual.

Although the orientation toward human-centered design stems from long-standing social-psychological concerns, the upsurge in computerized systems and the greater flexibility afforded by software have promoted the growth of these ideas.

In addition, the last half of the 20th century saw increased interest in the individual as represented by minorities, women, and so on. The concept of system design being oriented to the user can, from social, cultural, and historic standpoints, be an expression of an expanded concern for the individual.

This tendency has been accentuated by the growth of interest in rapid prototyping, which essentially hands over to the user (although not completely by any means) control over the analysis, design, and testing of design options. If information becomes the central problem of automatized systems and if information (having a behavioral basis) requires a user to decide whether the information provided actually is information (i.e., is the information provided actually useful?), this may help explain (although not totally) the increased concern for the user in design. The concern for the user deemphasizes the rationalist traditional approach while introducing new and even more vexing design problems.

It should not be thought that the traditional rationalistic design approach avoided consideration of the user. Had the strict Tayloristic interpretation of design held (personnel are productive units devoted to moving X tons of fertilizer from A to B in an 8-hour day), there would be no HF today because the essence of HF is concern for user needs. The question is where the concern lies. In the traditional HF concept, design was to be tailored to human limitations (i.e., thresholds) so that the operator would not be asked to do more than he or she could reasonably be expected to do. The user-centered approach accepts this, but considers it insufficient. The emphasis now is not on avoiding excessive demands on the human, but on exploiting human capabilities (e.g., enhancing operator pleasure in performing a task). This has also led to concern for mental models because in computer software the software design speaks at least indirectly to the concept structure of the user.

Because in computerized systems the operator often does not deal directly with the target system, but rather with a software program that controls the target system, the importance of communication between the designer and the user, and particularly in guiding the users to navigate the software, is increased. Because of its nature, software design becomes more individual and social than one finds in hardware design. Hence, there is a shift from the rigidly rational to the less rational.

Of course, there are so few studies of how designers actually design that we cannot know how great the shift from the rational to the social is or whether old (rational) and new (individual) design elements mingle.

Human-centered design represents a highly desirable goal. Who would not wish to make operation of a process control system much like playing a video game? It is right to recognize that the great bulk of people are like children in their obsession with games. Witness the frenzy over the Internet.

Assuming the desirability of making system design resemble a game (the authors are not Puritans who insist that work must be taxing), the question arises as to how one can incorporate these human elements into design. Traditional HF design has done so by attempting to eliminate discrepancies between equipment demands and human limitations. This is essentially a negative orientation: Do not require the operator to respond in 200 ms; do not expose the operator to excessive noise or vibration; do not demand the operator to have the cognitive sophistication that he or she lacks — popularly expressed, you shall not require your operator to have a PhD in electrical engineering to operate this equipment.

These efforts to ameliorate excessive system demands on users (a staple of traditional HF design) focused on elimination of negative features. If one takes the program of socially centered design seriously, it becomes necessary not only to eliminate road blocks for the operator, but — much more difficult — to provide positive elements in the design. How does one make the design of a machine fulfill operator aspirations? The machine should, if the reasoning is followed to its ultimate logical conclusion, become amusing, stimulating, cheerful, and so on. The problem the designer has with this is how are these features to be incorporated into design?

The problem of HF design has always been one of transformation (Meister, 1999) from a behavioral requirement to its appropriate physical expression in machine form. Up to now, the transformation limits could be accommodated in design, procedural complexities could be simplified, and complex displays could be redesigned. To make an equipment used in a job more like a video game is a much more daunting prospect.

The problem with theoretical ideals like those of human-centered design is that they must be reified; they must be made concrete. The difficulty is that designers do not know how to do this, and specialists do not know how to advise them. Ideals and goals are fine as long as they do not have to be implemented; then trouble arises.

Much depends on what the nature of the system is and whose design one is concerned with. The system permits the consideration of user concerns (i.e., allows or does not allow their incorporation into the design).

Much that has been written about such topics as socially centered design, the incorporation of flexibility, and so on in the design is admirable. One would wish to be able to include these factors, but the question is: How does one do so? How does one create a software program that is socially conscious? The opportunity to include the user in design depends on the degrees of freedom one has to design. For example, if the user interface is inherently highly proceduralized, the opportunity to introduce user considerations is very restricted.

The dichotomies implicit in theoretical writings (resulting perhaps from overenthusiasm) about user- and social-centered design (traditional design does it this way, user-centered design does it in a contrasting way) are misleading because these contrasts assume that the situation is either/ or. In actual design, it is not. One includes the user to the extent that system parameters permit and the system mission demands. The rationalist approach emphasizes the system mission to which all else is subordinated. The nontraditional approach (while not rejecting the importance of the mission) reduces its importance.

One question the reader should keep in mind is the extent to which theoretical concepts such as the preceding can actually influence real-world design. Design has its own imperatives and may be immune to theoretical concepts.

SYSTEM CHARACTERISTICS AND THE DESIGN PROCESS

In the previous section, we discussed system development as a general context within which each individual design problem is attacked. The individual design problem is much influenced by the nature of the system being designed.

There are, of course, differences in scale between designing for a single equipment or work station and a suite involving a number of individual equipments tied together by a single purpose or mission. Beyond that, there are differences among military, industrial, and commercial systems; between systems dedicated to a single customer and those sold to the general public; among systems that are mostly hardware or software or a combination of both; and between systems that are highly determinate or highly indeterminate (see Meister, 1991).

The type of system for which one is designing may have implications for the extent to which one must consider cognitive and information-processing stereotypes. Information is, of course, processed in all systems, but systems that are specialized for information processing (e.g., information warfare) require somewhat more consideration of information.

Adversary-type systems (military and highly competitive commercial systems like advertising) require consideration in the analysis design phase of what the potential enemy capabilities are to counter these in design. The capabilities and limitations of users for all systems require analysis, but more so in equipment with a variety of users and the impact of individual differences in users on operability and hence sales. Hardware and software systems require somewhat different design guidelines. Systems that have a determinate mission require different design provisions than those for systems having multiple and highly cognitive missions.

Design has always responded to technology, of course, but it seems as if technological change is occurring more quickly as time passes. One sees this vividly in the development of computer programs and graphics. The impact of this on the development of displays indicating the external and internal states of the system cannot be overestimated. Where equipments function in a suite (system), it is possible to transfer information graphically from one work station to another, to extract specific aspects of a piece of information represented on one console and expand those aspects on another console, and so on. The determination of what information the operator needs, how much should be presented, and how it should be displayed all become a matter of overwhelming concern for the behavioral designer.

Technological changes not only change the nature of the equipment/system, but also the role of the human in that equipment/system. Automation is the primary attribute to be considered because other attributes stem from it. In the automated system, the operator's role changes from an equipment controller (energizer, mover) to that of a monitor of an equipment whose equipment operations are determined by a software program. The operator may initially energize the system, but after that the program kicks in and the operator now need only keep a sharp eye on the displays to ensure that the equipment is not exceeding preset values. Displays are needed not only to indicate the system's progress in the performance of its functions, but also to be diagnostic of what has gone wrong if something does go wrong (as it inevitably does considering the propensity to programmer errors like the Y2K bug). These displays must make it possible to intervene before a catastrophic failure occurs.

One of the most important system factors affecting design is the complexity of the system being designed. *Complexity* is a difficult concept to define, and little thought has been given to it (except in reliability engineering). However, it can be grossly and physically defined in terms of the number of components involved in the system, the number of their interactions, and the number of equipment functions that must be performed.

Complexity also has behavioral connotations. Meister (1996a, 1996b) defined complexity in terms of the number of informational states that the system presented to the user during its operation. The user then had to interpret these states to operate the system effectively.

Complexity, therefore, has two aspects or faces. The first is an internal, physical one consisting of such components as circuits, which are not normally evident to the user and, correspondingly, behavioral complexity represented by stimuli displayed through a human–computer interface (gauges, dials, a computer terminal, etc.). Physical internal complexity determines behavioral complexity. However, because physical complexity is usually hidden from the user's view (except when it is revealed inadvertently, as in poorly designed interfaces), the specialist is only concerned with behavioral complexity. The engineer is, of course, responsible for physical complexity. However, if an analysis of internal functioning indicates to the specialist that the resultant behavioral complexity may be too great for the operator, the specialist must intrude on the designer's responsibility by warning him or her of the potential difficulty. Complexity on the internal level is the maintenance technician's problem.

Complexity can be scaled or at least its variations can be physically perceived. At the most molecular level, there is the single dichotomous signal: on or off, red or green, and so on. A higher level of complexity is information presented quantitatively by a single display (e.g., an altimeter). A still greater level of complexity is a continuing stream of signals presented separately to the operator; a still higher level is when the signal stream must be integrated and interpreted by the operator. Parallel to the stimulus aspect of complexity is a response aspect—what the user must do in response to the stimuli (e.g., a single response, turn the switch on or off; make a sequence of independent control responses; or coordinate these responses with the stimulus stream). At a still higher level of response complexity, the operator may be asked not merely to be aware of the signals, but also to interpret their meaning (i.e., to transform them symbolically and conceptually and then take action on that interpretation). On a gross physical continuum, complexity may be manifested in a single control or display, as a control panel or work station, as a total equipment, as a combination of equipments or a subsystem, and ultimately as a total system.

Complexity, therefore, has a number of manifestations: physical-external (the work station, the subsystem); physical-internal (number of components and their interactions); physical-perceptual (independent stimuli, combined stimuli); and behavioral-responsive (functions required of the user in response to stimuli).

The reader may ask: What does this have to do with design? Complexity in its behavioral form is transformed into information. The equipment is preeminently an information source and transmitter. With increasing complexity, the user must assimilate an increasing amount of information, which may strain his or her resources. It is an arguable hypothesis that increasing complexity is more responsible for inadequate operator performance than specific design errors, which can usually be avoided by application of design guidelines. One example is the PC user's need to know what procedures must be performed so that new information can be called up. The specialist must ensure that the designer keeps the amount of information and the way that information is presented within viable constraints so that the user is not (as is too often the case) overwhelmed by the mass of information presented. This is most likely in computerized systems, where there is a more direct liaison between internal physical system states and the user's conceptual structure.

Complexity also determines (and reflects) to a considerable extent what the system is supposed to do. The more complex the system, the more functions must (usually) be performed and the greater the strain on the user. Notice that these concepts fit within the S–O–R paradigm: The physical characteristics of the system present themselves as stimuli (S) for the user, who must perceive and comprehend them (O) and respond (R) appropriately.

Complexity has another effect on the system. At a very low level of complexity (e.g., the individual gauge, the thermostat), the device has only a function, not a mission. A function is a requirement built into a device; it either performs that function or it has malfunctioned. At a certain level of complexity, the system is given a mission by the designer, which is goal-oriented and requires a user to accomplish. As soon as there is a goal, a mission, and a user, the complexities pile up because the likelihood of human variability now increases. It is true that there is a user for a scale or control, but the number of interactions or rather the ways in which the user can interact with the control or display are very limited. For example, the user interacts only by perceiving the scale reading. An individual equipment, such as a lathe, does not have a mission, although it may have a number of functions. Only at a certain level of complexity (which is difficult to specify objectively) does a mission develop, and the human in the mission will affect that mission through certain mechanisms.

To design for performance of a function is much less difficult than to design for a mission. A component may have several functions, but in the performance of each function it has only a dichotomous state: It either performs that function or it does not. The mission may, however, be performed in a multiplicity of states because the human is at the heart of that mission. One can reduce the number of mission states by giving the human no autonomy: He or she can respond correctly in only one way, and any other way is failure. However, complex missions, even when performed on a purely physical level (e.g., broken field running in a football match), permit, and even demand, some variation of response.

In aiding the system designer, the specialist has a more difficult job when the system has a mission and that mission is complex. The effect of system complexity on design is therefore to incorporate in the design a mission that involves a multiplicity of behavioral states, each of which must be accommodated by the design. The analysis of complexity is therefore not merely a theoretical exercise; it is entirely possible to design a system that is so complex that average users cannot operate it.

At very high levels of complexity, the system develops or is caused to develop certain attributes. These include transparency or the extent to which internal system operations are revealed to the operator; and autonomy or provision for alternative modes of performing the mission.

The importance of complexity for specialists (because we cannot expect the system designer to be aware of the behavioral correlates of complexity) is that to do the specialist job properly a considerable amount of information must be acquired, analyzed, and interpreted. At a very low level of complexity, exemplified for this discussion by the individual scale or display, the role of the specialist is to eliminate negatives (i.e., to ensure that the design does not bump up against an inherent human threshold). For example, distances between scale marks must not be too narrow to be discriminated. If symbols are to be presented electronically, the size of the symbol (its font) must be large enough to be discriminated. Eliminating such threshold negatives is comparatively easy because they reflect black–white contrasts (either a symbol is large enough to be perceived or it is not).

At the mission level, there are relatively few absolutes. Here the specialist's job is to determine what the operator/user is supposed to do and, further, to imagine the various ways in what is to be done can be done. The designer suggests certain possibilities, some of which are less advantageous for the user. The specialist must show the designer that these are less effective. In most cases, their disadvantages are a matter of excessive complexity.

To recapitulate, complexity is manifested in information; that information about the system must be acquired, analyzed, and reflected in the form of a mission statement and a statement of user requirements. Then the mission statement must be analyzed to derive hypothesized structures or mechanisms that enable the mission to be performed successfully. The analysis of the design problem thus begins with the determination of what the user is supposed to do — the mission statement, which determines user requirements. In almost all cases, that statement as initially presented to the designer/specialist is inadequate. Equipment-oriented managers may transmit their design requirements with a complete statement of desired equipment performances, but the equivalent set of behavioral requirements may be completely missing or at least woefully weak. Therefore, user-centered design, like all design, begins by attempting to find out what the operator/user is supposed to do with the new system. It is only after user requirements or the mission statement have been fleshed out that the designer/specialist can start thinking about the mechanisms that allow the user requirements/mission statement to be implemented.

The concern for complexity is manifested most concretely when the human–computer interface is being designed because this is how information is communicated between the equipment and the operator; the operator responds by doing nothing (the system performs satisfactorily), noting a possible malfunction based on inappropriate information, or responding to a definite malfunction by manipulating controls (seeking to remedy the problem).

All possible information about an equipment or the world around one cannot be presented to the operator at the interface: There is too much of it and the operator would be overloaded. The engineering of physical sensors to represent equipment or world status would be too costly and too difficult. Complexity must then be considered when the designer examines component interactions to determine which of these should be represented at the interface. It is at this point that the HF specialist can make his or her greatest contribution to design by determining the amount of information — and only that amount — needed at the interface.

Our concept of the individual design solution process can be encapsulated in three words: *analysis*, *design*, and *test*. Of course, it is not as simple as that. For one, the test phase (and these are presumably sequential phases, although they overlap, they reiterate, etc.) is broken down into what we call *evaluation*, which is informal testing (involving such procedures as a walkthrough), and *testing*, which is formal (i.e., highly controlled), although experimental and control groups are not necessarily used.

In any event, the paradigm can be expressed in more detail as follows:

Phase	*Questions to Be Answered*
Analysis (the specific types of analysis, especially as they relate to software design, are described in Chapter 3).	What information is needed for each type of analysis?
	What information is available about the system and the user? Where can it be secured? ⬇
	What inferences can be made from this information? ⬇
Design	Development of hypotheses (design options) based on available information; ⬇
Test Evaluation	Which design most conforms to standards and design principles? ⬇
	Which design options are best (prototyping)? ⬇
Test	Based on formal test procedures (operational system testing, usability testing, experimental operational use), does the completed system satisfy behavioral requirements? ⬇
	If not, what must be changed?

The analysis–design–test paradigm is assumed to be what we call a *shell* or a conceptualization of a superordinate process within which each individual design problem can be solved. It is assumed — admittedly without proof — that all the nontraditional approaches to design can be accommodated under this umbrella concept.

In this concept, information is transformed by the creative and experiential resources of the designer into physical mechanisms. This transformation also involves decisions: how much the mechanisms will be automated and, in the case of computerized systems, how much sharing, if any, there should be between the computer and the operator/user.

For the test phase of the design paradigm, criteria must be developed by the specialist. These may be implicit, which means that they function without control, or they can be explicitly formalized. If the rapid prototyping process is employed, responsibility for criteria development is partially transferred to the user. To the extent that user performance is subjective (e.g., preferences, opinion), the evaluative criteria are also subjective and may have to be teased out by the specialist in user interrogations.

If information is as important to design as we say it is, then a major factor impacting design is the amount of information available at the start of design and how much more must be winkled out. As was pointed out previously, the major difference between initial, update, and redesign efforts is the amount of information immediately available to the specialist and designer.

2

The Design Process

In Chapter 2, we discuss design as a *process*, a term which suggests that activities are performed over time in response to changing stimulus situations. The process requires the application of methods to implement that process, but in this chapter we merely mention them when appropriate; more detailed consideration of methods are left to Chapter 3.

In this chapter, we describe a high-level model of the design/development/testing process (henceforth called, for summary purposes, *design*), which is organized around the questions needed to be answered by the collection and analysis of information or data.

A number of models (to be described later) have been developed to describe the design process; the purpose of these models is, we surmise, to exert control over that process. It is ironic that our entire technological civilization depends (at least in part) on the efficiency of that process and the systems, equipment, and products (system) it produces. Superficially that process is under no intelligent control except that of entrepreneurs and design personnel. The design process is ultimately a creative one, akin to painting and music; learning about its structure, as in painting and music, may help make it more efficient as well as place it under some control. Design process models may also help organize empirical studies of that process — studies that have so far been very few.

The questions to be asked are at first rather abstract and general, but become progressively more detailed as design becomes more molecular and detailed. The same questions are repeated in the various design phases, but in changed form because the level of detail to which the question refers changes the context of the original question and this slightly changes the question.

The design process is subject to verification by various types of tests of potential solutions and modification of these, based on test results that feed forward to influence later design decisions or backward to modify previous ones. The two major functions of the design process relate to functionality (which has physical consequences) and operability (which has behavioral ones). These two interact, of course.

Our model assumes both individual, team, and organizational behaviors. It is not a closed loop model; inputs from external sources (e.g., those relating to money and political interests) can strongly affect the design process.

Our model also assumes that design tasks are defined and prioritized. Task prioritization ultimately depends on perceptions by the design team of the parameters of what presents itself as a problem. Design is a problem because of the many unknowns in the situation. One starts with a specification of what the ultimate system should be (e.g., what it should do, how it should do it). These requirements are often obscure because they are incomplete and poorly phrased, and because the means to satisfy these requirements are, at least initially, obscure. There are a number of possible means to implement the design, but their advantages and disadvantages are, again initially, unknown. The entire design process is an effort to secure information that will narrow down the alternative means to a relative few and eventually one. Nothing about all this is fixed in concrete, nor is the process entirely logical because it depends on the skill of the design team members, their awareness of relevant parameters, and the comprehensiveness of their past experiences, which is often manifested in biases that may or may not fit the actual parameters of the design problem. Until the end of a process that may last months and even years, the design decisions made during this time are still problematic — until the effort is completed by verification testing.

Therefore, any design process model is a behavioral model involving basic human functions such as attention, perception, information processing, and decision making. Values enter into the model (e.g., satisficing, or the selection of a design that satisfies minimal requirements rather than one that is optimal but too demanding in terms of cost, etc.). The model assumes that there are different types of design problems, varying from those that are relatively well known (involving familiar parameters and solutions) to those about which little is known, and thus require more intensive information collection and decision making. The type of design problem is a major factor in determining how the model functions in engineering reality.

One can view our design model as one in which uncertainty is a primary factor and in which the design process is an effort to overcome that uncertainty. Design is an attempt to transform this rational process into a physical form — the system — that emerges finally like a butterfly from a chrysalis. Because everything depends on the design team, the process is, despite its physical engineering context, almost entirely psychological, and therefore a fit topic for behavioral study.

The design process has become more complex for the HF specialist because of recent efforts to humanize the system through a movement called *user-* or *human-centered design*. This is an effort to incorporate human qualities, human needs, and desires into the system. The desirability of doing this is not a question (systems exist for the human); the question is: How can this be done?

The user-centered approach in design theorizing (e.g., Norman & Draper, 1986) does not distort our model. Rather, it emphasizes the need for data about the user and, more particularly, data to permit transformation of behavioral mechanisms into physical ones. Our model envisages design as a highly rational, problem-solving process into which, because of the inadequacies of human reason, irrationalities in the form of intellectual biases and intuition enter.

For the behavioral specialist primarily, but to a lesser extent also for engineering, design represents a transformation of the behavioral (e.g., human thresholds and human stereotypes) into a physical form. If all design consisted of circuit boards, one would not need be concerned about the human in the design. However, in almost all cases, systems must be operated, maintained, and/or used by humans, and one cannot design without considering them.

Because the reader may encounter certain terms (and the themes these terms represent), they should be listed: *models, process, problem, detail, complexity, transformation, information, methods, personnel, decision making,* and so on. The design process we describe (however inadequately) in this chapter and book is one of incredible complexity, about which very little is known empirically, although a fair amount of theorizing — see the next section — has been conducted.

THEORIES OF DESIGN

Rational Theories

Theories of design and development are only theories. It is possible to put design into a controlled test situation, even an experimental one, as the senior author did many years ago (Meister, 1971), but this rarely happens.

It is possible to study design through observation, as did McCracken (1990), but actual system development is almost impossible to address experimentally (although it is possible to develop simulations and use them as experimental stimuli with design personnel).

Design theories are an attempt to understand extremely complex processes, many of which are covert. Designers may be aware of and understand only parts of that process. Of course, one can ask a designer what he or she is thinking about or doing and their subjective responses are artifacts, as is the documentation (storyboards, blueprints, memoranda) they produce.

Questions about design are fascinating:

1. Is design a purely rational process or does it have nonrational, perhaps chaotic, elements in it? (Answer: Of course it has such elements, but what are they, where do they come from, and to what extent do they influence design?) Because design is goal-directed, we deduce that conduct of the design process must also be rational. However, the goal may be rational, but the means of achieving that goal may be less so.

2. Traditional wisdom posits a top–down, logical (if–then) process, but suggestions have been made that a bottom–up process also exists. How does one define this process? To what extent does it occur, and do both occur concurrently? How important is each?

3. Macroergonomists like Hendrick (1997) assume that the overall organization has major effects on design/development. How much effect does the overall organization, which functions at a top level, have on a much more molecular design?

Rouse and Boff (1987) suggested that "organizational attributes can strongly affect the practicality . . . of alternative approaches to enhancing the use of technical information in design" (p. 12). Then they asked to what extent do organizational effects depend on type of design, and do designers perceive these effects as inhibitory?

4. To what extent should users be involved in design? Participatory ergonomics theory suggests that major user involvement in design will produce more effective design. True or false?

These are certainly not all the questions that arise in the course of studying the design process. For example, what is the effect of information availability (or nonavailability) on design solutions? How is needed information collected? How does the personality of design team members affect the process?

Obviously one can design without ever thinking about these questions, and it is likely that most designers think about them only sporadically, if at all. Design personnel respond primarily to the immediate problem. Why, then, is it important to develop and test theories of design as process? The answer is to secure some measure of control over a vast enterprise by understanding it.

Sage (1987) indicated that, regardless of any rubric one applies to the design process, the latter always involves:

1. Formulation of the design problem: The needs and objectives of the client are identified and potential design solutions are generated.
2. The design alternatives are evaluated.
3. These alternatives are compared in terms of the effect of each alternative on human and system performance results, and the most acceptable is selected.

This rational, top–down model (one phase follows another logically) is merely descriptive; it says what should occur, but not how the three processes are performed.

Sage pointed out that the rational process depends on information availability and that the types of information needed may change in the various phases. For example, in the initial problem-formulation phase, the information is used in a somewhat intuitive manner. In the later evaluation, comparative, and selection phases, the need changes to a more analytical type of information.

Lack of necessary information may cause the designer to fixate on a less appropriate solution, which is progressively refined. Individual items of information are combined to develop a more reasonable solution.

Design is a form of decision making, and decision-making theory is pertinent here; but it is decision making in a very special context. How one decides is a matter of familiarity with design problem criteria, the significance of the decision, and the time available to decide.

It has been suggested that there may be subconscious mechanisms in design. It is not clear whether these mechanisms are rational, based on experience. They may be as logical as conscious thought or random trial and error.

If the rational model depends on information, there are many ways in which one can consider information attributes. Sage listed accuracy, precision, completeness, sufficiency, understandability, relevance, reliability, redundancy, consistency, uncertainty, and so on. He defined information at various levels: (a) technically as related to information transmission over a channel; (b) semantically as related to information meaning and message efficiency; and (c) pragmatically in terms of its effectiveness in accomplishing the design purpose. The latter two are the most important.

Whatever information-utilization problems the engineer may have, the HF specialist has another problem. In addition to all the difficulties cited so far, the specialist has to transform behavioral principles and data into their physical equivalents. Information is one thing for the equipment designer who does not have to cross domain lines to make use of that information. Physical constants for bridge building, for example, remain physical in the process of designing the bridge. For the specialist, however, the behavioral/physical domain boundary must always be crossed. What does a certain anthropometric fact mean in terms of the pilot's visibility in an aircraft cockpit, for example? What does a particular cognitive stereotype mean in terms of developing a menu structure? The major difficulty in using human performance data is not that the data are incorrect or incomplete (although they often are), but that they must be invested with meaning for use in the physical domain; this is a profound transformation, which is quintessentially the problem of information application. This has significance for HF research, but behavioral researchers have stoutly resisted application over the years (see Meister, 1997).

Intuitive Theories

Another theoretical viewpoint sees intuition as critical to design. This seems reasonable if we view design as having creative elements. The problem is to define what that creativity consists of: If we oppose it to the rational theory, it seems as if creativity must be essentially irrational.

The answer may be that design involves if–then logic, in which the chain of reasoning between the *if* and the *then* is unconscious to the designer. Smith (1987) talked about knowledge *chunks*, which presumably contain the if–then logic within themselves. However, because these are holistic entities, the processes leading from if to then are submerged within the chunk and are therefore not visible (conscious) to the designer.

If intuitive design exists, it may tend to lessen reliance on overt knowledge sources and place greater reliance on leaps of faith. The knowledge resources available to HF specialists are meager enough without professionals using intuition as a rationale for failing to access more knowledge; lack of relevant data encourages intuition.

Leifer (1987) discussed design literature in terms of three categories. The earliest design material came from the Arts and is typically concerned with behavior of individual designers and the taxonomy of design methods and strategies. A second body of design literature comes from engineering and deals with creativity, optimization, and basic design principles. A third strand comes from computer science and, more narrowly, artificial intelligence (AI). This last category focuses on formal models of design, use of knowledge in design language, search, and human–computer interactions.

The lesson that Leifer derived from all this is that a number of different processes are involved in design so that the behavior of the individual designer may appear to be somewhat chaotic.

Obviously one can be a designer without being a theorist of design. To do so, however, means that to a certain extent one's design activity becomes unconscious to the individual and hence less effective.

Factors Affecting Design

The more conscious the design process is, the more likely it is that design errors will be avoided or intercepted. Certain factors may produce difficulties:

1. The design requirement that established the basis for the specific design project can be vague (the intended use or mission of the system is obscure) or may shift over time, as when the customer gradually seeks to add bells and whistles to the original design concept.

2. The customer may not specify the criteria (performance requirements, cost, reliability, etc.) to which design is performed and against which it is evaluated. An example of this is the customer saying, "I'll know it when I see it."

3. The designer's style (logical, intuitive, determined by experience, etc.) influences the kind of design solution that finally emerges from the design process. Successful design solutions of the past tend to be repeated, but these may not be appropriate for the present project.

4. The organizational constraints (e.g., time, money, number of personnel) have been alluded to already. Another constraint is the specification that a certain kind of equipment (e.g., a particular sensor from a previous system) must be used in the new design.

5. A major factor affecting design is the designer's mental model of the system being designed (what the designer knows about the system he or she is creating). This is both individual and communal (team model). The designer's mental model is one of the outputs of the design process; it is as factual as scenarios and storyboards, although entirely nonphysical. Initially, that model is incomplete and tentative. The initial inadequacy of the model may reflect and contribute to the initial inadequacy of the system design.

As the design develops, so do the individual and communal models. The HF specialist also develops such a model, but unless he or she is also involved in engineering details, that model is likely to be less detailed than that of the engineer-designer. Eventually, when the system is finished and given to the operator, the latter, too, develops a mental model of the system, corresponding, one hopes, to that of the designer. All these mental models constantly change during design.

There are, in fact, several kinds of mental model: a general experiential model within the designer that guides the solution to all his or her design problems; a more specific but still general model that applies to the class or type of system the designer ordinarily designs (e.g., a type of copier); and a more specific model of the system under design, which is a product of ongoing design. The first two models already exist when the designer begins work; the specific system model develops out of the immediate design problem, but there is an interaction among all of these. The development of the specific system model feeds back to the general design model and incrementally affects that general model. Changes in the latter are felt in later design projects.

Because design is usually conducted in a team context, one can ask, is a design decision a consensus decision? How is design affected by changes in the way the team develops? This occurs in a series of stages. The first stage is *forming* (i.e., the composition of the team is determined; introductions are made; and goals, scopes, schedules, etc. are laid out). The second stage is called *storming* (i.e., each member makes sure that his or her interest within the group is not compromised). The third stage is *norming* (i.e., members agree on mutual goals and their respective contributions). The fourth stage, and the one that interests us most, is *performing* (i.e., the team's work begins to be performed).

Another way to look at design is, as Ballay (1987) does, as a construction task—something concrete must be developed. What is important is how well the construction task is defined. Ill-defined tasks are so because not enough information is provided to specify the final product. Designers may create the information they lack by constructing prototypes and exposing these to potential users, in the course of which more information is gained, which finally enables the designer to construct an acceptable product. Prototyping can, therefore, be conceptualized as a effort to secure information.

Design Demands on User Behavior

The rational design process asks the designer to consider the effects of the design on the user, during the time the design is being constructed, and this requires complex conceptual manipulations by the designer.

It is assumed (by the behavioral specialist, if not by the engineer designer) that every design feature imposes a demand on the user's behavioral resources. In advanced systems, these demands are largely imposed perceptually, through displays, although in systems that require some operator control, like aircraft or driving an automobile, these demands may also be imposed motorically.

Changes in technology, now occurring rapidly, produce changes in the behavioral/physical transformation process; these changes usually impose more stressful demands on user resources. In an ideal design process, therefore, it is not sufficient for designers to seek technical solutions to the design problem; they must also seek solutions to (or at least be aware of) behavioral problems created by their technology. Advances in technology and their effects on behavior are seen most obviously in the design of computer graphics and multivariate displays. Displays that integrate information from disparate sources and display that information in a single display may not require different perceptual-cognitive processes than when information is presented by a number of discrete single-parameter displays. The likelihood is that human perceptual-cognitive processes are unchanging in their structure, but changing graphic presentations may produce unprecedented demands on these behavioral structures. The speed with which the new displays present data, the amount of information they present, and the integration they require of the user may challenge these fundamental structures.

People whose perceptual-cognitive structures are unsuited to the demands created by the changing displays (e.g., the elderly) may show reduced performance. Unfortunately, HF specialists lack knowledge of how to adjust design so that these stresses are minimized.

The problem of technology demand on user behavioral resources is a problem primarily for the behavioral specialist. However, from what one can see of the behavioral literature, specialists are not provided with the information they need. Paradoxically, when behavioral specialists introduce this factor into design, it complicates the design process.

Changes in technology do not make design less rational, but methods such as rapid prototyping, which accompany these changes, come close to a general cut-and-try strategy, which is inherently less rational because it substitutes unstructured user preferences for more rigidly structured designer processes. If, in connection with general consumer products, one attempts to incorporate consumer desires into the design, the process becomes much more intuitive because it means the designer must interpret/ transform a desire (which is always less coherent, less structured) into a physical product, which must take account of rational constraints. This also increases the difficulty of the behavioral/physical transformation process. Prototyping may be an effective way to discover factors overlooked by the design team, but the results of prototyping should not change the design object or scope.

Alternative Approaches to System Design

A number of alternative, nontraditional approaches to system design have been developed.

1. Sociotechnological system development, of which macroergonomics is a special form. It evolved from English studies at the Tavistock Institute and emphasizes the matching of social and technological systems and the environment (Pasmore, 1988). Czaja (1997a) indicated that the design principles associated with the sociotechnological emphasis are often vague and tend to overemphasize the social aspects of the use system while ignoring the technological aspects.

There are aspects of the sociotechnological approach that are relatively objective. The system is viewed in terms of inputs and outputs and the resources required to transform one to the other. This is obviously one way in which one can analyze the system, but the input–output relationship does not necessarily translate into design parameters.

2. Participatory ergonomics, which emphasizes the role of the user in system design. The goal is to utilize the special knowledge that users possess and to include their needs and desires in the design process. The methodology makes use of focus groups and quality circles.

3. User- or human-centered design (Norman & Draper, 1986), which has been described previously, is not much different from participatory ergonomics.

4. Computer-based system design emphasizes the use of computer tools (computer-assisted design [CAD]) to solve design problems. It is not a total design process, but is a method of helping to solve design problems.

5. Ecological interface design (EID; Rasmussen & Vicente, 1989) represents an extension of skills, rules, knowledge (SRK) theory (Rasmussen, 1983) to the design of how information should be presented to system personnel. As such, its use has been focused on very advanced display design, but Vicente (1999) has said that it describes total system design.

6. An extension beyond user-centered design is something called *socially centered design* (Stanney et al., 1997). The basic assumption of this approach is that work occurs in a social setting; by observing how people work on a daily basis in specified work settings, it is possible to develop more meaningful system design requirements. How one gets from work settings to design requirements is not very clear.

The one thing that all these approaches have in common is that they attempt to move beyond what Stanney et al. called the *traditional Tayloristic system-centered design,* which they defined in terms of the phases of problem definition, generation of system alternatives, analysis of these alternatives, and selection of the one best design. (See Sage, 1987, for an identical approach.) Stanney et al. called the approach described in this book as traditional — lacking concern for the people elements in the system. The authors of this book disagree, but neither side can convince the other.

The concern for the human in the system is not really new. The approaches all grow out of the Tavistock studies, and these go back to the early 1960s, if not earlier. The system approach (Koffka, 1935; Van Gigch, 1974) certainly mandates concern for the human in the system. It is fair to say that human-centered approaches have had limited success in actual design, although experimental studies do report some advantages. One problem with these approaches is that they emphasize only one aspect of the overall design problem.

The Concept of the System

The preceding section suggested that how one conceptualizes the system one is designing at least in part determines *how* that system is designed. Following Lind (1988), we conceive of the system in terms of what he called the *goal-seeking approach.* That approach stems from general elements used in system analysis: (a) a goal or set of goals; (b) alternative means (system elements or strategies) that can be used to reach these goals; (c) resources that the system possesses and are expended during its efforts to achieve the goal(s); (d) a set of relationships (Lind called this a *model*) between the goal(s), the alternative means, and the requirements imposed on resources; and (e) a criterion by which the preferred alternatives are selected.

Manifestly, we are dealing here with the same rational approach to system design as that of Sage (1987). The reader may recognize that these categories are similar to previously described stages of system design. Lind's model, which is taken directly from operations research and engineering principles, can be translated directly into those design stages.

The design process we describe in this book is most efficient when it models itself on the system concept (but the modeling is largely unconscious because designers rely on the logic inherent in design rather than on theory).

Theorists will ask, where in all this is the user, the environment, the social context? All of these modify the system elements and are taken into consideration to a greater or lesser extent as the designer values these. The utility of the theories is that they attempt to bring concerns to the designer's attention. For example, one can think of the user as one of the resource elements (e.g., in the sense of providing information as in prototyping) that can be used to modify design. The same could be said (to a lesser extent, however) of the environment and the social context. Certainly one of the system goals could be user satisfaction or social facilitation, provided the customer wishes to make these one of the system goals.

The system has multiple goals; the enrichment of the human and his or her context can be, if one wishes, one of these goals. However, the human cannot, except in special circumstances, be the only goal (except possibly for amusement devices).

One way to flesh out the system development skeleton is something called an *abstraction hierarchy*, which owes its genesis to Rasmussen (1986), although there have been further refinements in recent years (Vicente, 1999). The purpose of abstraction in the goal-seeking approach is to identify means–ends relationships between system elements. These relationships correspond to fundamental analytic processes in system design, as can be seen from Fig. 2.1.

What Fig. 2.1 represents is the decomposition process, which, as pointed out earlier, is fundamental to the design process. The abstraction hierarchy depicted in Fig. 2.1 is equivalent to the progressive detailing of design. Thus, we can describe the system in terms of its goals arranged according to goal–subgoal relations. We can further describe the system in terms of its functions, which also can be ordered according to function–subfunction relations. Finally, we can describe the system in terms of its physical structure, which can also be ordered in terms of whole–parts

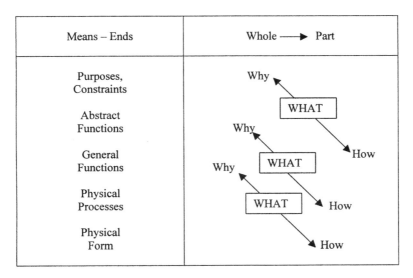

FIG. 2.1. The abstraction hierarchy (why, what, and how; Rasmussen, Pejtersen, & Goodstein, 1994).

relations. The three elements of the hierarchy (goals, functions, structures) represent three modes of decomposition.

The two dimensions of the abstraction hierarchy (means–ends, whole–parts) are independent of each other, which means that a particular function can be implemented in different ways. This is the basis for the determination of alternative means to implement functions. From this, the necessity of selecting among the alternatives arises.

The Context of System Design

The system concept and system design processes postulated here are based completely on logical reasoning premises. However, when we look at design as it is actually pursued, we see frequent variations. Where do these come from?

The factor responsible for these variations is context. The basic design process is maintained, but it is adjusted and modified by the individual design situation in which it occurs. To understand any specific design situation, one must examine its context. The factors providing context variations include, first, the nature of the design problem, which demands more of certain kinds of activities and less of others. Design differs depending on whether one is dealing with hardware or software; whether the design problem is one of original design, update, or revision; whether design involves a single work station, a complex of work stations, or a complete system; whether the design is a military or commercial program and whether the latter has a restricted number of users, like pilots, or a general consumer; what the time scale for development is, how much money is available, and how many personnel can be put on the project; who the intended users of the system under design will be (ordinary adults, children, skilled, unskilled, disabled); and whether the designer is working largely alone or with one or two colleagues on a design team or is part of a large-scale project. One can ring a great many changes on the alternative possible situations.

The one constant in this picture is that design (if one uses the term to characterize all the activities involving in creating a machine) is highly variable, responding to a great number of contingencies.

In all of this shifting uncertainty, there are certain immutabilities. First, regardless of whether the problem is simple or complex, the design team is always faced with a problem. Design is always performed within certain constraints and has certain knowledge resources. Design is always a matter of analysis, evaluation, selection, and testing. The design output is never completely satisfactory to everyone because compromises (satisficing) must be made.

DESIGN AS INFORMATION PROCESSING

Introduction

Design is a matter of making choices among possible goal and goal-implementing alternatives. Choices demand information unless the choice is completely random, and we know from preceding pages that design is (or attempts to be) a coldly rational process.

Information is neither secured nor meaningful unless one has a prior question that the information answers. So everything begins with questions, and the remainder of the chapter is organized around the questions listed in Table 2.1.

A few words about these questions. We say that the questions do not change substantially as we proceed from one design phase to another, although the more detailed context in which the question is asked modifies that question somewhat. Design proceeds from the molar to the molecular; at first, initial design analysis is concerned with relative abstractions, like goals and functions, and then proceeds (still within the initial design analysis) to a consideration of alternative implementing possibilities. For example, one might input data to a computer by voice, keyboard, or writing on the terminal face. These are analyzed, and a selection of one of the alternatives is made.

In detailed design and evaluation, the breakout in Fig. 2.1 of functions to subfunctions to subsubfunctions (the decomposition process) begins and now most of the previous questions asked earlier are asked again. Not all questions, possibly. For instance, if the design team is satisfied with the system design goals it has set, it may not ask questions about goals, although perhaps about subgoals.

There is no sharp demarcation between time-adjacent phases (no party to celebrate moving from one phase to another). Questions asked in one phase may not be completely answered until one is fully in another phase. The effective designer is one who periodically reviews the questions asked previously and considers whether their answers are pertinent to the new phase. The designer's situation awareness (i.e., his or her understanding of the variety of events occurring) determines, in part, that design's effectiveness.

In all of these phases, products are produced as part of the design. One major product is concepts (how does one design to implement a particular function) and physical structures (e.g., circuits, modules, software code). These are evaluated according to criteria first developed in initial design analysis and revised when necessary.

TABLE 2.1
Major Questions Asked During Design

Initial Design Analysis

---

1. What is/are the goal(s) of the system to be designed?
2. What is the scope of the design problem (e.g., is the design new, an update, or a revision? How difficult is the design problem?)?
3. What type of system is involved (e.g., aircraft, automobile, word processor, general-purpose commercial product)?
4. What parameters are involved in the design problem? What questions about the details of the design problem must be answered?
5. Are all system requirements fully described in the design specification? What behavioral implications (e.g., threshold limitations) can be drawn from these requirements?
6. What further information is needed to make design decisions? Where can that information be found? What technology and information can be transferred from previous related systems?
7. Who is the operator/user? How important is the human role in system operation? What is the human in the system expected to do? Does the user have any special qualities, needs, or desires that must be taken into account? Have personnel requirements been specified in design requirement documents? Can they be developed?
8. What factors (e.g., technological requirements and constraints, environmental conditions, etc.) may affect personnel functioning? What nontechnical factors (e.g., money and time restrictions, political aspects) affect overall design and design of the human role?
9. What design tools (e.g., analytic, evaluational) are needed to support the design effort?

Detailed Design and Evaluation

(All prior questions apply in addition to the following):
10. What are all the possible ways in which the system can be designed? How should the human–machine interface be configured? How adequate is the interface design?
11. What additional information must be collected and analyzed as a result of increasingly detailed design? Where can this information be found?
12. What criteria can/should be applied to decide among the various alternatives?
13. Of all possible design situations, which is the most effective? What human performances are required at progressively more detailed design levels?

Design Testing and Verification

---

14. How should operational system testing (OST) to verify the adequacy of the final SEP be conducted? What criteria and test methods must be specified? What compromises must be made among competing criteria and design factors?
15. How adequately do personnel perform in the OST? What last-minute fixes must be made to the production model of the system?

We distinguish between evaluation and testing. The former involves no user performance. Evaluation includes modeling and simulation (to be described later). Design evaluation is much more common in the first two development stages (initial design analysis and detailed design) than is testing, which is more common toward the end of the design cycle. At least this was true before rapid prototyping came along in the wake of computerization. Prototyping has elements of both evaluation (in terms of user preferences) and testing (when users are asked to exercise the proto- type). Prototype testing is somewhat informal (how otherwise could it be rapid?), but in the final design phase (design testing and verification), which is coordinate with the start of initial production, testing is (or should be) quite formal and as highly controlled as a laboratory experiment.

Initial Design Analysis

The starting point of the design analysis is a question: What is the system that is to be designed (i.e., what is its goal)? The goal has certain attributes (implications) that need to be investigated: A specified speed requires an engine of a certain horsepower, voice recognition poses its problems rela- tive to the range of voices the system recognizes, the reliability of that rec- ognition is a question, and so on. A critical feature of all design is that ev- erything has consequences (implications) and that those consequences must be investigated, even if, in the end, they are found to be unimportant.

The nature of the design problem (its scope) must be considered. Just how difficult is the problem? This is influenced by the fact that the design is either new, an update, or a revision. Of course, a design is almost never completely new because in almost all cases it is a linear descendent of a class of systems.

Again, this is a problem of information. What does the predecessor sys- tem have to tell the designer? That predecessor provides a context in which new design proceeds. An update is a matter of building onto an ac- cepted design structure, and a revision points to the nature of what must be revised.

Another item of information also helps scope the problem: the nature of the system being designed. A traditional equipment like an automobile is designed and redesigned in increments over the years. There is no major change however much the automobile companies tout their products. To design a commercial space plane that will carry a thousand or more pas- sengers at the periphery of space obviously requires much more innova- tive design and is consequently much more difficult than developing a new model (update) of an automobile.

The scope of the problem requires the specification of the parameters involved. For designers of a new aircraft, these may be such factors as type of engine and its thrust, cruising altitude, speed to be achieved, and aero- dynamic qualities. For the specialist who is concerned with a cockpit, one parameter may be the degree of automaticity to be included (the glass cockpit; Wiener & Nagel, 1988), the type and number of stimuli to be presented to the pilot, the required speed of response to enemy missiles, and so on.

Each parameter raises questions that must be answered. If the aircraft is highly computerized, how much control should/can be left to the pilot? Studies (Amalberti, 1999) have shown that overautomation may, under certain circumstances, degrade the operator's performance. If the aircraft is a fighter, it is necessary to think of possible enemy threats (e.g., air-to-air missiles), and this raises the question of what warning displays are to be provided and how rapidly the pilot can perceive and interpret the displayed information and respond. Every question produces a need for relevant knowledge that is either easily available or must be sought.

If design logic uses a deductive paradigm (if–then), implications are the essence of this paradigm. If the customer imposes a certain requirement, what are the consequences (implications) of that requirement? If the designer specifies a certain design feature, what consequences will flow from that feature? In almost all circumstances, these consequences demand further information, which sets off a further cycle of information seeking.

The implications of the system requirement for the human may be of several types, but all involve stressing the individual. There is the necessity for considering the physical and physiological limitations of the human (e.g., visual and auditory thresholds) and, more important, cognitive limitations (e.g., the inability to deal with more than 7 ± 2 items; G. Miller 1956), certain preferred ways of handling information, such as how displays are to be interpreted and individual items of information are to be correlated. All of these thresholds have difficulty as the root cause, and stress is what they produce. These implications may be merely registered as possibilities in the very early analysis phase; later they may be considered deliberately in an effort to measure them, at least by scaling, if not more quantitatively. The trouble, of course, is that outside of physical/physiological threshold limits, almost nothing has been done with perceptual and cognitive thresholds, which become much more important in computerized, information-processing systems.

Because all questions imply the need for information, this leads to other questions: Where is the information to be found? What information and technology can be transferred from a predecessor to the new system? This is a matter of identifying common parameters and functions and determining whether these commonalities can be transferred in the context of the requirements and constraints of the new system.

Many designers may not exert themselves in this search for information. If the information is not readily available to them, they may suppress the need for it and continue to function in a state of uncertainty. They may also hypothesize an answer to a question and then assume it is correct.

A most critical question for the behavioral specialist is: Who is the user and/or operator of the system to be designed? More important, what is this human supposed to do? The behavioral specialist must be aware of the physical functions to be performed by the system (because these determine the behavioral functions the human performs), but his or her primary concern is the derivation of behavioral functions from physical ones. For example, if a display is presented to the operator of a system, the behavioral functions of monitoring, analyzing, and interpreting are inevitably involved. The more information provided by the technology, the more detailed is the derivation of the human function/subfunction. If the designer of an interface merely says a display will be provided, the specialist can deduce only that perceptual-cognitive activity is required. If the designer says that the display is a multivariate display reporting the interactions of multiple dimensions (altitude, speed, attitude, etc.), the specialist may be able to decompose the perceptual-cognitive function into more meaningful subfunctions.

In the if–then design logic we referred to previously, technology provides the *if* and the specialist provides the *then*. In examining the effect of technology on human functions, it is necessary to make a distinction between operators and users. The distinction is exemplified by the differences between the pilot of a commercial airplane and its passengers. Pilots are active in controlling the aircraft, whereas passengers are ordinarily passive unless there is an emergency. The range of behavioral factors one must be concerned with in the case of pilots is, therefore, much greater than that for passengers. One cannot ignore users because they also have their needs, but the pilot's needs are much greater. There is an overlap between operators and users because all operators are also, under certain circumstances, users; certain basic functions do overlap (e.g., as in lavatory facilities and seating space).

From a behavioral standpoint, therefore, certain dimensions inform behavioral functions: control versus lack of control, activity versus passivity. The importance of the human role in the system is critical. An operator performing many functions that cannot easily be automated commands much more attention than a situation in which there is only minor human involvement.

Another important factor is whether the design specification has imposed any personnel requirements that need to be implemented. In the authors' experience, few if any design specifications include such requirements. Therefore, in most cases, the real needs of an operator are a matter of inference rather than of specification, which leaves it up to the behavioral specialist to extract those needs, their implications, and the functions involved.

The specialist should consider all factors (e.g., technological, environmental, etc.) that may impact human functioning. The nontechnological factors (i.e., time and money availability) that impact the design problem must also be considered by the specialist. In situations in which time is at a premium, how important is a detailed task analysis (TA) of the human in the system? In a rational design universe, choices must also be made among/between alternative methods of analysis based on relative utility. What can one reasonably do with restricted time and money resources?

User- and social-centered theories of design aim to increase the human influence on design solutions. This inevitably brings up the question of how far one can include user influence in design. What specifics of the user must/should be included and how?

The usual behavioral analysis of a design problem takes the human into account by attempting to determine what the human is to do and what the implications of those actions (in interaction with relevant technology) are. Designers, being rational beings, do not wish to design in such a way that operators are inordinately stressed by what they are expected to do. They rely on the behavioral specialist to tell them when human bounds are exceeded. When they make errors in this connection, it is because they (and the specialist) lack necessary information.

This is a negative activity — the avoidance of harmful situations. What the extreme theorists of the human want to do is include positive aspects of human functioning in design. The basic problem is that it is often quite unclear how one can do this. In a life-threatening situation for a fighter pilot, how can one include worker satisfaction? Is satisfaction the same for all workers? How can one even define *satisfaction* in technological terms when that satisfaction must be transformed into physical mechanisms?

One can do this for devices designed specifically to provide users pleasure, like computer games. The device has no function or mission other than to provide satisfaction. There is an artistic element here that cannot easily be provided in devices that have a mission purpose (other than pleasure or satisfaction).

Moreover, there is a cost associated with determining what the user wants. Prototyping, which includes the determination of user preferences, costs money, time, and personnel resources. There is the cost of testing, analysis of data, and development of physical mechanisms that respond to human desires. Realistically, company management will not pay for this unless the system under design is specifically for amusement. We are left in most cases with efforts to avoid negative design features, but this has always been an HF responsibility.

The need to secure certain information to answer certain questions is what animates the selection of a particular method (and design as a whole as well). The method selected is the simplest and easiest to perform that will provide the necessary information. Zipf's (1965) law operates throughout design: The least amount of effort and time needed for the design team to answer a question drives design.

Detailed Design and Evaluation

This phase begins when the outlines of a general design solution have been developed. Although the name *initial design analysis* suggests only an analytic process, it actually includes, or may include, the entire range of activities associated with design, including development and evaluation of alternative configuration possibilities to the point of drawing blueprints and writing software code, testing these perhaps by prototyping, and selecting a single preferred configuration, which must in later design phases be fleshed out. The selection of a single, general configuration was based on a range of possible criteria from an inchoate *gut feeling* to some form of value weighting, which quantifies designer judgments concerning each of the possible alternatives (Meister, 1985a).

Among the criteria for technological advantage are the following: Which configuration uses presently available technology to greatest advantage? Which configuration can make most use of off-the-shelf components? Which configuration is simplest, most reliable, and least costly? Configurations that push the state of the art (e.g., require four computers vs. one) are less preferred. Tried and true design solutions that have been tested by experience are preferred to novel ones. Operability plays a much lesser role among selection criteria, except in a negative manner: Any configuration placing an obviously unacceptable stress on human thresholds must be rejected or at least altered to avoid the threshold problem. Simplicity is also an operability criterion (fewer actions and decisions required of the human, simpler displays). The more decisions required of an operator, the greater the opportunity for error.

Much of the same activity as in initial design analysis proceeds in this phase, but at a much more molecular level (e.g., modules, components, circuits, the writing of implementing software code, etc.). The same questions are asked again that were asked previously, but they are subtly different because they are asked about a more detailed configuration. The range of alternative implementing means is narrower in this phase. For example, if a circuit is needed to perform a function decided on in initial design analysis, the number of ways in which one can design a circuit is limited. If data must be input to a computer, the number of ways one can effectively input data is restricted (presently to the keyboard, voice and touch-screen technology).

Do specialists and designers ask these questions self-consciously as questions? The likelihood is that these issues are not far from their minds regardless of whether they articulate them.

The behavioral specialist continues in this phase to examine the system technology being designed to ensure that it does not contain elements negative to the human. However, as the components become more molecular, it is likely that they become less transparent (visible to the operator/ user). Hence, in many if not most cases, the human has no role vis-à-vis those components.

This is not the case with the human–machine interface. The interface (displays and controls) extracts certain formerly opaque (to the operator) components and functions and makes them visible via displays.

The specialist's concern for the interface involves the following questions: (a) What information about system and external world status is needed to ensure mission success? (b) What are the alternative ways in which the interface can be configured? (c) Which interface configuration will permit the operator to do his or her job most effectively?

Prototyping, in the sense of exposing users to alternative configurations, may in fact be more important in interface development than in other design aspects.

Design Testing and Verification

The phase we describe in this section is the culmination of the design cycle. As such, it is relatively short; one can think of it as the end of detailed design.

To this point, any behavioral testing that has occurred has been with units other than the complete system (e.g., modules of the total system). The culmination of this phase, which is coordinate with the start of production and precedes system turnover to the customer, is testing and verification that the system can be used to achieve system goals.

The questions raised in this phase are: Can the aircraft actually fly 2,000 mph? Will its terrain-following radar actually enable it to find a target 1,500 miles from the start point? Can personnel detect a subsurface target reliably? Many of these questions are solely physical ones (e.g., will fuel consumption enable the aircraft to fly 2,000 miles without refueling?). Other questions are combined physical and behavioral ones (e.g., will the new sonar apparatus permit detection by sonarmen of "super-quiet" vessels?). There may also be questions of a completely behavioral nature (e.g., how well do navigators read coordinates on a new electronic map system?).

Testing to this point has given the design team only partial answers to the question of system adequacy because the final system has not yet been exposed in its complete form to the total range of conditions to which the operational system is routinely exposed.

From a behavioral standpoint, system verification testing is conducted to answer a number of questions:

1. Can representative users exercise the system well enough that original design goals are achieved?

2. How well (quantitatively) can personnel perform with the new system?

3. Assuming that certain performance discrepancies are found, can/should these be fixed and how?

It is necessary for the design team to review the original design goals described in the design specification. Over design time, these may have been modified if initial analysis and tests have shown that the original goals could not be accomplished as originally specified. For example, the aircraft customer may have originally desired an aircraft that would fly 2,000 miles without refueling, but design experience may have shown that only 1,500 miles could be achieved. Hence, the design goal had to be changed. It is possible that design is littered with such compromises; these must be examined anew to ensure their validity and to help designers to remember what they are testing.

The design goals also need to be concrete by the criteria specification. If an aircraft has been designed to fly to a target 2,000 miles away, how much error in finding that target is permissible: 1 mile, 2 miles, 5 miles? If an underwater target is to be detected by a new sonar array, must the target always be picked up (100% reliability) or is 95% satisfactory? If a stamping press was designed to be operated with an error of no more than .0005 millimeters, will .0008 be acceptable? Each system goal should have some sort of quantitative criterion associated with it. In many cases, the criterion must be deduced at the time of testing because it was only implied originally.

The criteria we are talking about are quantitative; qualitative judgments of performance adequacy are not acceptable unless the performance being measured is inherently qualitative (e.g., aircraft handling qualities, like the Cooper–Harper Scale, 1969). In those cases in which quantitative criteria cannot be developed, designers should examine the system goal because, in the absence of quantitative measurement, the goal may be meaningless. The quantitative criterion may require special instrumentation to be measured; if there is no money for instrumentation, a qualitative alternative may be utilized.

Questions 1 and 2 asked earlier—can goals be achieved and how well can personnel perform—are not quite the same. It is possible to determine that the ultimate system goal has been achieved without knowing what the actual human performance is that achieved that goal. How much one wants to learn about system performance is a judgment call because it is obvious that, just as in a task analysis, there are levels of task description, so there may be levels of detail in the test data. For each measurement, the question that the measure illuminates or answers should be specified. Measurement for the sake of measurement is not as acceptable as art for art's sake.

The decomposition process gives us subgoals and subfunctions, tasks and subtasks. Test personnel must decide how far down in the decomposition process they want to measure. Goals, certainly, but do they wish to measure at the subgoal, subfunction, and subtask levels? The more one measures, the more information one gains; but how relevant is the information and how costly?

Ideally, one would wish to determine performance at all possible system levels, but this may prove to be impractical. There is a distinction between testing at the operational level versus testing at the technical level. For example, at the technical level, the type of hatch lever may be of concern; at the operational level, the concern is, how fast can one open the hatch? Perhaps one should measure only those behavioral performances that could affect overall system performance. Thus, one would not take physiological measures of the operator because physiological factors are not likely to determine system performance (except under strenuous environmental conditions like extreme cold and heat).

Measurement at a subfunction level can be useful in diagnosing why a particular performance deviates from the goal. For example, if users exhibit only partial satisfaction with a product, it is useful to know how accepting they are of its packaging, its color, the ease of opening it, and so on. This kind of information is very valuable if it becomes necessary to fix something that measurement has shown to be inadequate. The same thing is true if the system has a mission; if the goal is not adequately achieved, it may be because a subgoal's performance was responsible. Only a detailed examination of the test record would reveal this.

The development of an effective verification test procedure is therefore a fairly complex one (see Meister, 1985a, 1991). What is needed to be known are: (a) what system goals, subgoals, and so on are tested, (b) what criteria and measures are applied to the measurement, (c) what the conditions are under which the test is conducted, and (d) what answers one should expect from the testing.

The answers to these questions are especially important for the behavioral specialist. Engineers are attuned to sophisticated measurement of equipment, but they have a tendency to assume that the physical measurement subsumes the behavioral one. For example, because an operator has to throw a switch to energize an engine, the measurement of engine thrust accounts for operator performance. In other words, engineers may feel that human performance is purely instrumental in exercising a physical equipment; thus, if the equipment works, this automatically verifies the adequacy of the behavioral performance.

This argument might have some validity if there is only one way in which the operator can perform. In cognitively driven systems, the operator has much more flexibility, and a particular system output may not be clearly understood until one knows how the operator made his or her input.

The development of a verification test procedure ordinarily requires the writing of a formal document to be signed off by engineering management. It is assumed that the human performance aspects of that test are under the control of the behavioral specialist and therefore are written by him or her.

Performance of the test should involve the following conditions because a valid test is not possible without them:

1. The total system is to be tested — not a system lacking essential modules or a system that has been tweaked by technicians to perform at a higher performance level than would ordinarily be manifested. If the equipment being tested is in some ways incomplete, this may be a partial performance test, but it is not a system test.

2. The conditions under which the system is to be tested are those of the operational environment (OE). This means (a) determining what OE conditions of operation are, and (b) incorporating those conditions into the test. Outside of running the test in the OE, which is in many cases not practical, it is necessary to replicate its conditions in what amounts to a simulation. Ideally, this would occur in a very realistic full mission sophisticated simulator, but these are ordinarily available only for complex, expensive hardware systems.

Therefore, the specialist must attempt to analyze what the OE is as it relates to the system under test. For complex situations, this is not as easy to do as it might seem. Our knowledge of the behavioral aspects of the OE is seriously lacking primarily because the OE has never been considered by HF specialists as a topic for research. At a minimum, test conditions must include the requirement that, once the system has started its mission, there must be no unrealistic interference with its performance, not even if the system fails or otherwise performs inadequately (the tendency for management to send in a squad of expert technicians as soon as a malfunction occurs must be resisted because it is unlikely that the operational system will have such facilities immediately available to it).

Another requirement is that test subjects should be representative of the ultimate users. This may not be a problem if the system is a general-purpose commercial item. However, if the user population is restricted to those who have special training or skills, test subjects must possess that training or skill or be given special instruction. The natural desire to use design engineers as test subjects to make the system *look good* must be avoided.

Behavioral specialists have a choice of test methods (see Meister, 1985a; O'Brien & Charlton, 1996) because it may be necessary for them to accept substitutes. The ideal mode of measurement is often too expensive in terms of money and time.

The verification test is not a laboratory experiment, but if it is to be considered valid, it must be conducted in standard fashion. The test plan must include a specification of what statistical tests are to be performed and what answers are expected from such tests.

The Verification Usability Test

We define a verification usability test not in terms of performance of subsystem tests in a laboratory during development (see Nielsen, 1997), but in terms of giving a production model of the system to a customer/user for the latter to use the system as if it had been purchased. The customer/user relates its experiences and preferences, and so on. Usability testing in this sense is like prototyping, but with the complete production model of the system. An example is beta testing of new software (Sweetland, 1988), in which selected potential customers are provided free of charge with a program to use, after which they are expected to report deficiencies to the developers.

Verification usability testing can be meaningful only if certain controls are imposed by the developer. These include:

1. A description of the way in which the system is actually used by the customer; possibly a diary should be compiled even if automatic recording is available.
2. A record of all system malfunctions, their derivation, and what was done to correct the malfunctions.
3. Measurement of operator performance with the new device to the extent made possible by technology (e.g., automatic logging of control manipulations).
4. A judgment of customer/user opinion about the performance of the new device recorded on scales that have been pretested and refined.

The costs associated with such a verification usability test are probably less than those involved in a formal operational system test (OST). If sufficient control can be exercised over such a usability test, this is a reasonable alternative to the OST.

The Fix

The OST and verification test are assumed to be the end point of the design cycle; after such a test, the system is handed over to the customer who then accepts all further responsibility for it.

However, there is one small preliminary. It is commonplace for any final test to reveal defects, deficiencies, and inadequacies that should be fixed before the production model is turned over to the customer. Such deficiencies in the physical structure take center stage and are modified as necessary.

Because of the assumption that physical performance subsumes behavioral performance, behavioral deficiencies are more difficult to get fixed than physical ones. Unless a performance deficiency seems to have an obvious effect on overall system performance, it is likely to be overlooked or ignored by engineering management. If such a defect can be easily remedied by, for example, a change in the operating procedure, the change is made routinely. If the change requires some modification of hardware or an appreciable amount of software code, engineering management may well dig in its heels and resist the change. If a change can be made inexpensively and quickly, fine; if it cannot, the importance of the deficiency is denigrated by managers and reference is made to what we call the *Horatio Alger* solution (i.e., the ability of users to compensate for the defect by willpower, experience, ingenuity, or, by that great standby of American engineering, special training to overcome the difficulty). The authors have seen homemade changes with wire that solved small problems quite well, but such solutions may not always work. The reluctance of engineering management to fix difficulties is well known.

3

Design Methods

In this chapter, we review in somewhat more detail than was possible in Chapter 2 the various methods that have been developed to aid the behavioral design process. It is important to know what underlies these methods; it is not enough to use them superficially simply because they exist. They do not provide desired answers unless the specialist specifically asks the questions described in Chapter 2.

This is not a complete description of these methods; that would require a separate volume. Moreover, these methods have received extensive treatments (e.g., Laughery & Laughery, 1987; Luczak, 1997; Meister, 1985a, 1991; Weimer, 1995). The reader who wishes more detail can consult these. The methods to be described in this chapter are listed in Table 3.1.

The specialist's primary analytic methods are function and task analysis (TA), which are so closely tied together that they are practically one. Function and task analysis derive from the decomposition process described in Chapter 2. We differentiate between function analysis and function allocation; the latter methodology is supposed to allocate functions between and in conjunction with the human and the machine.

A function is something that must be performed by the system or some part of the system. A function requires at least one and usually more tasks to complete the function.

With these definitions, the initial system concept can be analyzed by using function analysis to produce a list of functions to be performed. Those functions can be allocated to the human, the equipment (hardware), or the combination of human and equipment (plus software). For those

TABLE 3.1
Behavioral Design Methods

1. Training, experience, and logic.
2. Function analysis: determination of functions to be performed from a description of activity. Function allocation: assignment of task responsibility to human, machine, or both.
3. Task description: description of activity in stimulus–response terms. Task analysis: determination of factors influencing task performance (includes graphic tools such as operational sequence diagrams and function flow diagrams); these are also used in function analysis.
4. Design evaluation methods: checklists, walkthroughs (verbal performance simulation), and physical simulations.
5. Psychological research methods: performance measurement, interviews, questionnaires, scales, critical incident descriptions.
6. Rapid prototyping and usability testing: user analysis and exercise of initial design versions.
7. Cognitive engineering techniques: cognitive task analysis (CTA) and ecological interface design (EID).

functions allocated to the human, task descriptions must be composed and task analyses performed.

Task description (TD) and task analysis (TA) must be differentiated, the first describing the individual task in terms of stimulus and response elements, the latter describing tasks in combination and interrelationships that result in task groupings. TA, which is performed by questioning the task description (these questions are described later) permits the making of judgments about workload and stress effects, design implications, and skills and knowledges needed to perform tasks.

Underlying all the methods are the specialist's training, experience, and logic, which we consider fundamental to all the methods, but which have no formal step-by-step procedures.

One may ask whether training, experience, and logic are actually methods. They are, of course, factors that determine how well a specialist uses other, more formal methods. At the same time, they are methods in the sense that they are responsible for creating the questions the specialist asks about the system requirement, the tasks to be performed, and the design hypotheses that the designer creates. Certainly training, experience, and logic are methods when employed in the absence of other formal methods. For example, there is no formal method for drawing behavioral implications from a mission statement; the implications derive from the analyst's accumulated knowledge and the logic of the statement.

Most of the methods used in prototyping and other forms of testing have been carried over from experimental psychology. Therefore, they need no special description because it is assumed that the knowledgeable reader is familiar with them.

In the past few years, considerable attention has been paid to what are called *cognitive engineering* techniques (Woods, Watts, Graham, Kidwell, & Smith, 1996), with special attention to cognitive task analysis (CTA) and the method called *ecological interface design* (EID). The rise of the computer, because of its emphasis on information processing and as a consequence of assuming from the operator more primitive perceptual-motor responsibilities for running systems, has made cognitive functioning the centerpiece of the operator's work.

Although not widely employed, certain methods are of special interest because of their potential. Workload estimations, error analysis and prediction, and cognitive mapping are examples, as is human performance modeling.

Some of the more specialized function analysis methods were derived from industrial engineering and apply primarily to production processes. Because most design specialists do not deal with production, we do not discuss these. The reader who is interested in these should consult Laughery and Laughery (1987) and Luczak (1997).

There is a great deal of variation in methodological context. Some of these methods, like function and task analysis, are used more frequently than others; almost none is used in all design situations. Some of these methods are used in design analysis, others in testing; some have been around for many years, others are very recent newcomers. Some are in general use, others are largely experimental, while still under development and tryout. Needless to say, some of these methods are more useful than others.

The fact that certain design methods are described in the literature does not automatically ensure their use. Much depends on the nature of the design project, its scope, and the availability of time and money. Any method takes time to apply, even when computer software packages are available that reduce time and work effort. Consequently, there is a cost–benefit ratio involved because it is always possible to perform behavioral design analyses with few, if any, formal design methods. It is assumed that these methods produce better designs, but there is no empirical verification of this.

Rationally, a method is selected for use because it most readily and effectively answers one or more of the design questions listed in Chapter 2. However, nonrational factors enter in. For example, a customer — usually a governmental one — may mandate use of certain methods, in which case the methods are applied regardless of their value.

The availability of supporting instrumentation also partially determines selection of a method, particularly one that is computerized. For example, one would not select a computerized method unless one had a computer with the necessary capacity. Other factors, such as time and cost, influence method selection. For example, complex methods like human performance modeling, which require the development of a great deal of input data to make effective use of such models, may not ordinarily be selected. Idiosyncratic factors (e.g., the specialist's knowledge) in part determine whether a particular method is utilized. If the specialist is unaware of a particular design tool, he or she obviously is not able to make use of it. Presumably, also, specialists have preferences for methods based on prior experience, and these are used more frequently than others.

No method has an inherent solution to a design question; it is the specialist's interpretation of method outputs that provides answers to the questions. What the specialist brings to a method in terms of training, experience, and logic may be much more important than the method.

CRITERIA FOR SELECTING A DESIGN METHOD

The methods exist, but they are not used automatically. The question then becomes why designers select a particular method. Some methods, like function and task analysis, are so frequently used that it would be unusual if they were not used (i.e., they are traditional).

However, the primary criterion should be the need for a particular technique, and this of course ties in with the design questions one asks. If the specialist does not ask these questions, he or she probably does not make use of appropriate methods. Moreover, these questions and the methods used to answer them are not asked unless the human has a significant role to play in system operation. Unless that role is significant, and there are major unknowns about how the human performs in the system, little attention is given to behavioral concerns. That is an additional reason that one of the highest priorities for the specialist is to determine what the human role is. If the system under design is an update of a traditional one, like a new class of copier, it is difficult to attract the attention of system designers to the results of behavioral analyses because designers tend to believe that prior behavioral analyses do very well for the new system, especially if major subsystems are imported into the new system with little or no change.

Designers of the new system argue that the old system interfaces transferred from the previous system have been validated by use. Of course, the previous system may not have received proper attention, and users may have adapted to poor interface design.

If there are major unknowns that require behavioral analysis, the method selected must be able to provide reasonable answers to the questions asked without inordinate effort and in minimal terms.

Engineers use whatever they think works. This tendency is increased because design personnel access information sources in inverse ratio to the difficulty of securing the information. They much prefer to use their own knowledge resources and those of fellow workers rather than search for data (Allen, 1977).

The methods we describe can be thought of as essentially ways to secure information and are utilized in the same manner as any other information source. For example, prototyping is an effort to secure information about initial designs.

What is a reasonable answer to a design question is difficult to specify. If the question relates to a number of parameters, the answer must encompass those parameters. For example, if the question involves an automatic voice-recognition device, the answer must take into account auditory frequency and intensity variables, with potential environmental factors.

The level of specificity and precision of the answers provided by the selected method are also considered. The source is often a factor. For example, in producing a function analysis of a new type of aircrew system, intense one-on-one interviews with experienced pilots might be valued above a questionnaire sent to aircraft designers because designers have more confidence in face-to-face interviews.

Other factors that affect the selection of a design method include the stage of design. Items of information change their value according to the design phase in which the information becomes available. Because the methods are information sources, this also applies to them.

Therefore, the methods can be evaluated on the basis of utility and the amount of work required to use them. For example, if a method requires elaborate diagramming of tasks, it is less desirable than one not requiring such diagramming. Because the amount of information a method provides is not quantifiable and therefore not obvious, less demanding methods are preferred.

The specialist must have some preconception, some hypothesis, about what he or she gets from a particular method before selecting it. For example, one would not apply a workload estimation method unless one suspected that workload was going to be an important factor in the design. From that standpoint, it might be said that one of the things the design tool does for the specialist is to confirm certain expectations about the role of the human in the system.

LOGIC, TRAINING, AND EXPERIENCE: PSYCHOLOGICAL RESEARCH METHODS

Logic, training, and experience come into play as soon as the specialist begins to analyze the design specification and other design requirements documents because there is no formal methodology to derive behavioral implications from requirements. Throughout design, there are analytical/evaluational functions to be performed for which formal design tools have not been provided. For example, the specification of alternative design options and evaluation of design outputs like drawings are largely performed without formal design methods.

How many behavioral design specialists recognize the importance of semantics in their analyses? For example, task description and analysis are essentially semantic exercises when the task is described verbally. The decomposition of a task is identical to the way in which one was taught to parse a sentence — to break it up into noun, pronoun, verb, and object. For example, to decompose the task "operator throws switch" into stimulus–response terms is to do the same thing one did when analyzing the statement presented as an English sentence. More sophisticated analyses of the task are accomplished by adding adjectives and adverbs (e.g., the operator throws the switch slowly) or with a modifying phrase (e.g., after examining the display). It is also likely that analytic tools that employ graphics, like operational sequence diagrams (OSDs) are, after the initial perceptual scrutiny, translated into language equivalents before conclusions are derived from the graphics.

We have said that the specialist is likely to select a design tool based on prior successful experience with that tool. He or she begins to use the design method with a certain expectation of what he or she receives as outputs from the method. In some cases, the method is used to confirm or deny hypotheses generated before applying the method. Specialists may not express these hypotheses overtly, and they may not even realize that the hypothesis serves as the rationale for using the design tool.

If the specialist already knows the answer to a design question, what does he or she get from the method? Where graphics are part of the method (as in OSDs, function flow diagrams, decision/action [D/A] diagrams, time lines, etc.), the graphics enable the specialist to see relationships among system elements more clearly. The graphic aspect of the method accentuates and thus renders the relationships more visible.

Among the skills and knowledge the specialist brings to the evaluation of the design are those derived from graduate training in experimental psychology. These include knowledge of the principles of experimental research; development and use of subjective techniques like interviews, questionnaires, and scales; and statistical analysis procedures, as well as knowledge of computer procedures. All of these are fully exercised in prototyping and operational verification tests.

FUNCTION ANALYSIS

Introduction

A number of things need to be said about function analysis (FA) as a general methodology.

There are two stages in function analysis. The first is the determination of the nature of the function. The nature of the equipment to be designed (e.g., a radar set) or the activity to be performed specifies the functions required (in the case of the radar, detection and identification of the target, estimation of bearing and range, speed of closure) by the equipment and what one does with that equipment and activity. It is therefore quite easy to list the functions involved. Unfortunately, such a list of functions does not tell one very much.

The second stage of FA is the more important part of the process. The specialist must deduce from the nature of the functions and the modality through which the functions are expressed (in the radar example, vision) the implications of their parameters.

These implications specify the factors to be taken into account in these functions (e.g., the size and shape of target stimuli, the rate of signal return, background stimuli, etc.). These must be compared with the human's visual thresholds to answer the questions implied by these parameters: Will the operator be able to see target stimuli, discriminate them from background noise, and make necessary responses to the stimuli (detect and identify the target and calculate its bearing, range, and speed of approach)? Knowledge based on the research literature is necessary to answer these questions. Every function of every type of equipment has implications; the important thing about FA is its exploration.

Because it is the easier thing to do, we suspect that most specialists stop their analysis at the stage of listing the functions required by the type of equipment or activity. The listing of functions is easy, but the follow-up analysis is much more difficult. Consequently, there is a tendency for the specialist to leapfrog the stage of drawing implications from functions and to proceed directly to the task.

As far as function allocation is concerned, we do not know what designers actually do in practice. Obviously functions are allocated because we see this in subsequent design decisions, but we do not know whether designers actually conduct a session in which they decide (as in a Chinese restaurant) Function A to the human, Function B to the computer, and Function C to be a joint responsibility. This is certainly the end result of design deliberations, but any suggestion that a special process takes place apart from ongoing routine design activities appears unlikely. Engineers often allocate functions on the basis of having the equipment perform as many of the functions that are within, and even a little beyond, the engineering state of the art. All of the functions left over go to the human. In specific types of systems, because of the history of their development, many functions have by tradition been allocated to the human or machine, and the new design does not change this. With the new computer technology, the question may become not to which — equipment or human — shall this function be given, but in what way will joint custody of a function be allocated, and in what ways do the human and computer interact. This is a much more difficult and serious business than allocation between human and machine.

Theoretical discussions of function allocation periodically raise their fuzzy heads in the literature. The most recent (Sheridan, 1998) involved a debate about whether function allocation could be performed on a rational basis — yes or no.

Sheridan made a point which suggests that FA as a general methodology is largely a chimera. "Function and task are used here to mean the same thing, though some authors make subtle distinctions" (p. 20). Much of the theorizing is generated by semantic confusion. If by rationality one means deliberate conscious examination of alternatives, then function allocation is partly rational, involving the comparison of machine and human capabilities, and partly irrational because some of the decision making involved occurs just below the surface of the designer's mind. Not surprisingly, the discussants in the Sheridan debate agreed to disagree.

Functions and tasks are often confused. A function, which is much more general than a task, is ordinarily associated with a capability to do something; a task is something specifically required by a specific system design. Logically, the function should be considered first because the function requires the task to be implemented. However, because the function is often so general, in actual design practice one develops the specific task first and only then does one classify the function implied by the task. Designers often ignore the derivation of the functions, going directly to the task.

Function Analysis Tools

The statement of an individual function provides only limited information for the designer. If something is to be monitored, there must be an information presentation device; from this one can imagine a number of source possibilities (e.g., discrete lights and scales, a computer screen with alpha-numerics, etc.). The function should also suggest a number of questions to designers (e.g., what information, presented in what form, occurs in what time cycle, what must the operator do after receiving the information, etc.). The last bounds the function and translates it into tasks. The need to answer these questions (even in preliminary form) leads to information-collection actions by the design team.

Whether function identification is an important part of design depends on whether the new system is truly novel, in the sense of having few or no predecessors, either direct or as a member of a class of systems. A transportation system that depended on mental imaging of the location to which one transported oneself (a not infrequent science fiction theme) would demand intensive function identification and task description because one could not refer to practices involved in a predecessor.

FA tools become important only when one asks how functions interrelate over sequence and time. To assist in more readily visualizing these interrelationships, the tool adopts a graphic mode.

Flow Analysis.   Flow analysis is simply the sequencing of task operations from some start time to the completion of the task. The flow may be of events and information as well as of people from one site to another. For example, one could describe the pilgrimage of Abraham and his family from Ur of the Chaldees to the Holy Land in the form of a flow diagram. As long as there is a sequence of events, it can be described in a flow analysis.

Time, as such, is not necessarily described in the flow diagram. One can insert real-world time values into the diagram, but the important element is the sequencing and interrelationship of events. Such flow diagrams have been widely used in industrial processes from which they were derived to describe human performance events. A functional flow diagram for HF purposes is shown in Fig. 3.1. The flow diagram is used not only to

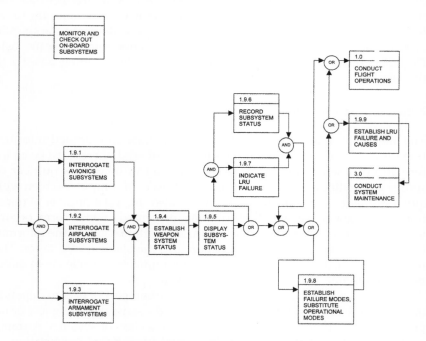

FIG. 3.1.   Sample functional flow diagram (modified from Geer, 1981).

describe a sequence of operations, but also to stimulate questions about whether a particular function is needed, can be combined with a similar function, or can be modified in some way to improve its performance. The flow chart in any of its manifestations (process chart, flow diagram, OSD) is entirely natural because humans think in terms of a sequence of events, and that is all a flow chart depicts.

Operational Sequence Diagrams (OSD). OSDs were developed by Kurke (1961). Their purpose was to graphically represent the information-decision sequence the system (including the operator) must undergo to accomplish a purpose or mission.

Kurke indicated three uses for the OSD: (a) to establish the sequence of operation requirements between subsystem interfaces at various levels of system analysis; (b) to evaluate control panel layout and workspace designs; and (c) to ensure that all system functions and their interactions are accounted for.

Although the central concept of an OSD is the decision, other processes are required to describe an operation. The symbols used in OSDs are shown in Fig. 3.2. Information is received, a decision is made, and an action is taken. A sample OSD is shown in Fig. 3.3.

Laughery and Laughery (1987) indicated that the process of developing flow analysis charts is useful because it forces the analyst to gather information, define functions, and arrange sequences. In other words, the process of charting forces the specialist to be systematic, logical, and analytical when he or she might ordinarily not be. The flow diagram is, therefore, a control mechanism.

Time-Focused Analysis. Timeline analysis (and the charts that result, although it might be more correct to say that the charts come first and analyses afterward) is an analytical technique representing the temporal loading for any combination of tasks. If flow or sequence is important to functional flow charts, time, particularly when concurrent operations are required, is the essential feature of timeline analysis. The underlying assumption of this analysis is that having to perform multiple actions at roughly the same time tends to exert an excessive demand on limited human resources (perceptual, cognitive, and motoric functional capabilities) and therefore may produce an unacceptable stress on the operator. The purpose of the timeline chart is to help define points at which such stress may occur and thus help ameliorate them. The technique as used by specialists in the course of design is to predict workload and scheduling.

Timeline charts are developed by adding time information to flow analysis techniques, particularly when actions must be performed concurrently. An example of such a chart is shown in Fig. 3.4.

SYMBOLOGY

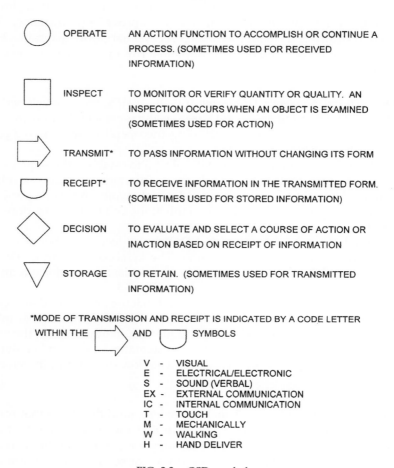

FIG. 3.2.   OSD symbols.

At a time in which less and less is done manually, there are computer programs that can input time data and produce workload profiles for whatever operation the specialist is interested in. Nevertheless, the computer requires inputs (in this case, time) before it can arrange those inputs in a meaningful way.

The timeline chart is utilized to estimate the amount of task loading on the operator, but this is a difficult problem because no one knows quantitatively what the effect on human resources is of having to perform concurrent tasks. One may ask, what good is a timeline chart if one cannot make such an assessment? However, one can set as one's goal the reduc-

SECOND-LEVEL FUNCTION 2.4.1 PERFORM
PRESTAGING CHECKOUT

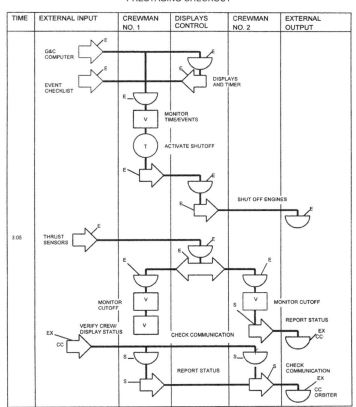

FIG. 3.3. Sample operational sequence diagram (from Geer, 1981).

tion of time stress, and the chart is useful in identifying points in the task sequence where these stresses may occur. Another possible use for timelines is to determine that the operator's best performance times are within the system's time allowance for task performance.

Link Analysis. This is a network-type analysis that has proved useful in dealing with layout and arrangement problems. One can use link analysis in a variety of ways, but its most frequent use in the past has been control panel and work station arrangement. The basis for the link analysis is the frequency of the interrelationships among the elements that one wishes to describe. This requires at least a gross task analysis: The sequence in which the actions interrelate is not as important as the fre-

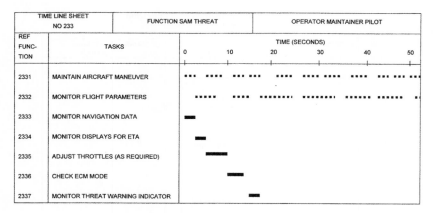

FIG. 3.4.   Part of a timeline sheet (from Geer, 1981).

quency with which they occur. The usual objective in the layout problem is to associate the elements that have the most frequent interrelationships so that, for example, two controls that are most often activated together should be physically co-located or two displays that are interrelated by function or importance or frequency of reference should be physically located together. The link analysis is a cut-and-try process; one develops a preliminary layout and analyzes it using link frequency. If one is not satisfied, one tries a different arrangement.

For all the preceding methods, there are computer programs that presumably reduce the amount of work involved. However, in each of them, certain input data are required, after which the program goes to work. In the case of optimizing a control/display layout, the analyst would first have to specify a proposed layout, after which the computer would analyze the link frequencies involved. The analyst would then modify the layout and ask the machine to make another cut at link frequency analysis. A sample link analysis is shown in Fig. 3.5.

Task Network Modeling.   Task network modeling is a technique that is described in more detail in the next section; it deals with human performance simulation models and stems directly from task analysis and its descendant graphic methods, such as the flow diagram and OSDs. Some computer programs such as Systems Analysis of Integrated Networks of Tasks (SAINT) are specifically adapted for such task network models.

In task network models, human performance is decomposed into a series of subtasks, and the relationships among these are described by a network. The network includes nodes that represent discrete subtasks performed by the operator. Sequence is important as in task analysis because the network structure defines the order in which personnel perform these

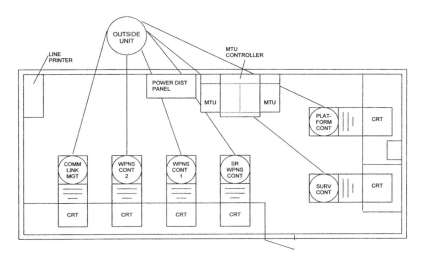

FIG. 3.5.   Sample link analysis for workspace layout (from Geer, 1981).

subtasks. Each node has associated with it certain pieces of information: (a) time to complete the subtasks; (b) if the subtask is always followed by the same subtasks, the subtasks to be performed after the present subtask is completed are noted; (c) if the subtasks may be followed by several subtasks, the list of possible subtasks to be performed after the present task is noted, as well as the way the next subtask is selected; (d) any user-defined relationship that defines the human's effect on the system or vice versa; and (e) an arbitrarily assigned subtask number and name.

Once these items of information are determined for each subtask, a computer model of the task network can be developed using a number of task network simulation models. Simulation experiments can then be performed by varying these items or indeed the network structure. For example, one could alter required time to perform one or more subtasks, change error likelihood, the effect of environmental stressors, and so on and see what the effect of these changes are on probability of successful mission accomplishment.

Task network models are most easily adapted to procedural tasks, but modeling languages can also accommodate hybrid models, in which less proceduralized functions like decision-making or perceptual-motor activities can be modeled using theoretical structures adapted to these less formalized functions. A sample network analysis diagram is shown in Fig. 3.6.

The specific function allocation that is decided on results from solution of the design problem, rather than from a conscious decision to allocate a function one way or another. Where such a specific decision is required, as

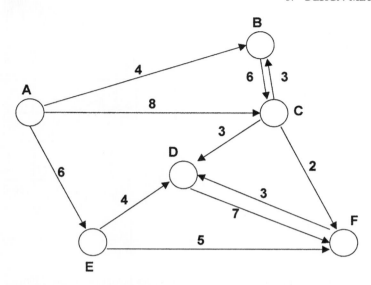

FIG. 3.6.   Sample network analysis. *Note:* Numbers represent the frequency
of contacts between nodes (e.g., work stations; arrows indicate which work
station initiated the contact).

between human and machine, it is because there are two or more possible
design solutions, not functions.

For example, it is assumed that a computer-controlled telephone
switching system occasionally malfunctions. It is technologically possible
to enable the computer to recognize deviant switch readings and perform
its own reset. This is complete automation without the need for human intervention. Of course, a human could perform this function.

Designing software that monitors its own operations and performs its
own corrective actions is tremendously expensive, so the system designers opt for the human function. The design solution was not determined
by a technologically advantageous function, but by a nontechnical factor
(cost).

If either the machine or human cannot perform a function, the other determines design. If both can perform, the human or machine alternative is
selected on the basis of technological advantage (e.g., greater speed, more
reliability, but, most important, lower cost).

The famous Fitts comparison of man versus machine (1951) was an effort to develop generalizations. The concept of comparing human and machine attributes is acceptable, but only within a specific design question.
The comparison cannot be used for arbitrary noncontextual decisions.

Any comparison of human versus machine requires knowledge of human and equipment capabilities. Too little is known specifically about the

former to permit a real comparison. Moreover, the function-allocation decision is not simply one of operation; it must also be based on maintenance concerns, which becomes more important when the system fails.

TASK DESCRIPTION AND ANALYSIS

Introduction

Although it is not common practice, we distinguish between task description (TD) and task analysis (TA). TD is the description of the individual task in stimulus–response terms: Tasks are listed but not interrelated nor analyzed. TA describes the individual tasks, but also interrelates them with each other in sequential (mission-determined) order and makes certain inferences (analyses) about the potential effect of the task and its characteristics on operator performance. For example, the specialist may examine the task to determine whether any task characteristics are likely to be excessively taxing or productive of error; he or she may bring these to the designer's attention with recommendations for task redesign.

Task analysis can be viewed as an attempt to make the behavioral design process more rational and less intuitive. It does so by establishing a set of analytic and graphic conventions that specialists must follow if they intend to make use of the method. It forces the specialist to become more orderly and systematic in his or her thinking.

It is possible to make behavioral judgments about design without performing a task analysis, but those judgments may then be less efficient. The important thing about the TA is the asking of questions about the behavioral aspects of the task.

Both FA and TA are concurrent activities usually begun in the initial design analysis phase, but FA is usually completed in this phase, whereas, depending on the nature of the design problem, TA may continue to be performed well into detailed design.

What does the specialist want the TA to tell him or her? It is assumed that the task makes demands on human resources (psychophysical, perceptual, motor, and cognitive). Therefore, it is of interest to anticipate and possibly take measures against any excessive stress caused by the nature of the task — stress that will reduce performance efficiency.

Interactive Task Factors

Task analysis varies in only a limited number of dimensions. The first is detail. This is the process of decomposition of behavior, which was already discussed. The same behavior can be viewed at progressively more detailed levels or ways of viewing. One can start by using the task as goal oriented: What is the operator attempting to do? One can then describe the task in terms of what the individual does to accomplish that goal; whether he or she is successful at it; if not, what decisions he or she must make; and so on. One can look at the task in molecular action terms such as those in time/motion analysis (e.g., "turn the door handle"). One can ultimately decompose any task-oriented behavior into physiological terms, but for design purposes this is usually unnecessary.

Task description can also vary in terms of what the design problem is. If the focus of what the analysis is concerned about is the decisions the operator must make, one develops a cognitive TA. The industrial ergonomist who is concerned with repetitive motor actions will emphasize these in his or her behavioral description.

Another dimension of TA is time. Two times are involved: total mission time (which is relatively unimportant for the analysis) and what one can call *descriptive time*, which is how finite the time units are into which one breaks the individual operator actions. One can make these as fine as one wishes. For example, at one extreme, time and motion study breaks these down into individual hand and eye motions.

How much system context one wishes to include in the task description depends on the analyst. Some theorists like Luczak (1997) wish to include the organization matrix in which human–machine operations are performed, and there is some point to this because that context does influence task performance, although how much is unknown.

Another factor of importance is the degree of dependency between tasks. Dependency is defined by the effect that Task A has on Task B if Task A is not performed or is performed inadequately. If, as a result of Task A effects, Task B cannot be performed or is performed less than adequately, one would say that Task B was highly dependent on Task A. The relationship may also be mutual if the tasks are performed concurrently.

What is most important is what the task analyst wants to do with the results. The analyst attempts to reduce the demand levied by the machine on human resources. Then the analyst needs to (a) determine the difficulty level of the task, and (b) the probability of error (although probably not in quantitative values).

Because a detailed TA depends in part on equipment details, in most cases the equipment configuration must precede a detailed TA. Hence, as far as the machine is concerned, the TA offers little to the designer.

A primary judgment of difficulty includes but may not be limited to the following indexes:

1. The operator is required to perform an action that is beyond threshold limits, such as responding to a stimulus faster than 200 ms;

2. Several operations must be performed concurrently by the operator;

3. A threshold (visual, auditory) is approached, although it is not exceeded;

4. The speed with which any single or multiple operation approaches a minimum threshold;

5. The machine/system requires the operator to perform highly complex cognitive analyses of (a) patterns of symptomology derived from multiple displays; (b) detection and monitoring of stimuli that are just above operator thresholds; and (c) important decisions to be made on the basis of incomplete and unverified information about obscure states of the system and of the real world;

6. The operator is required to monitor transitory states over long periods of time without overt feedback.

The reader can observe that there are several dimensions to difficulty: physiological thresholds; sequence (including concurrency) and speed, and what can be termed *cognitive complexity*, which refers to the demand imposed on cognitive faculties.

TA says least about this last and requires the most intuition of the analyst. Although cognitive complexity has not been studied extensively in the past (although see Hendrick, 1984), one can expect it to be on the increase; the computer makes this inevitable.

In some cases in which TA methods are used, this is done after the fact of design and to persuade designers to take some redesign action. TA can also be used throughout design to document what the operator must do.

TA is influenced by the development process. In initial design analysis, when the effort is aimed at making gross molar decisions, a detailed TA is not needed and may not be possible because the level of detail in a TA depends, to a certain extent, on details of equipment functioning. If the TA must describe a control movement by the operator, the control must be specified and is not specified until a certain level of detailed design has been reached.

Another use of the TA may be to help the designer understand what is involved in the operational functioning of the complex entities he or she has created. This is not merely a theoretical rationale. As systems become more complex, the designer may not be able to comprehend completely what has been created; this produces the Gestalt *emergent* (unexpected output) and is responsible for many inexplicable effects, which a new software system often presents. The TA unravels the complex threads of the task, and this helps the designer understand the ramifications of his or her design (provided, of course, one can get him or her to look at the TA).

Once a number of design options have been hypothesized, performing even a rudimentary (rough and ready) TA of each option can be useful in selecting one and in making choices. The problem in doing so is that the time frame available for hypothesizing and selecting design options may not permit even a rudimentary TA, even if the equipment information for each design option is available.

Once an initial design configuration has been selected, the development of a detailed TA, in conjunction with the more detailed design, is highly desirable to serve as a check on the details of the latter. It is entirely possible that some feature of the design has been overlooked, and the detailed TA points out this feature.

It is not conceptually difficult to develop a TA: The major problem is to secure the information needed to describe the operator's activities. Some of this information may be secured by making inferences from equipment flow diagrams. Additional information can be secured by interrogating equipment engineers about how they view the equipment functioning. Concurrency is important. If the TA is to be useful, one cannot wait until formal pieces of an engineering paper are published because these documents are often delayed while more important design events occur.

As a record of what the system and particularly what its operators are supposed to do, the TA as an archive needs to be continuously updated so that everyone on the design team knows where he or she is during system development. A major value of the TA may, in fact, be its function as a recording document. It can also be used as a primary source for the development of operating and (to a lesser extent) maintenance procedures.

In the testing phase of development, the TA can serve as a means to decide what data to record in the test and at what test stages. The TA also plays into the use of simulation techniques and models to be discussed later. Human performance simulation methods could not be developed or used without first applying a detailed task analysis. Rapid prototyping as a species of design also demands some form of TA.

A very large number of task aspects are listed in Table 3.2, which is based largely, but with some revisions and additions, on a taxonomy developed by Kirwan and Ainsworth (1992). The sheer number of categories in Table 3.2 suggests that even a simple task is a phenomenon of considerable complexity. One does not expect the specialist to take all of these into account. Table 3.2 serves primarily as a sort of checklist to remind the task analyst of what he or she should look for in the task analysis output. Many of these aspects may turn out to be irrelevant to a specific design problem, but the specialist should be aware of them.

Particular attention should be paid to the last category in Table 3.2: TA questions to be answered. This is what the specialist hopes to output from the TA. This last category is in the nature of a wish list; some of the subcat-

TABLE 3.2
Factors Describing and/or Affecting Task and Task Performance

Nature of the Task
  Decomposition into stimulus–response elements
  Level of task description detail
  Task function and purpose
  Sequence of tasks over mission time
  Description of stimuli initiating the task
  Nature of operator responses required: activation of controls; motor function; analysis
    of displayed information (perceptual/cognitive)
  Response speed and accuracy required
  Concurrent tasks and coordination requirements
  Nature and degree of task interdependence
  Task complexity and difficulty
  Amount of operator attention required
  Task criticality

Initiating Stimulus Factors
  Displayed information
  Information received from outside the system
  Communications within and from without the system
  Kinesthetic stimuli (e.g., handling qualities)
  Malfunction/emergency conditions
  Completion of previous mission stage

Human Resources Activated by Task
  Type of resource (e.g., perceptual, motor, cognitive)
  Extent to which resources are required

Hardware Features Affecting Task Performance
  Controls and displays involved and their characteristics
  Critical system status values displayed
  Geographic arrangement of controls and displays

Task Ouputs
  Changes in system status
  Feedback provided to operator

Possible Task Failure Consequences
  Errors of omission and commission can result in
    • Variation in system performance
    • Degradation of system performance
    • Complete malfunction of the system
    • Damage to the system

Task Analytic Questions to be Answered
  What is the probability of error for task and task sequences?
  What are the consequences of task error?
  What knowledges and skills are required of the operator?
  What factors in the task and the operational environment are stress-inducing?
  How much stress is there?
  Are any human thresholds exceeded by task requirements?
  Will the task produce feelings of excessive difficulty or the contrary (boredom and
    inattention)?
  What level of manning is required to perform system tasks?

egories cannot be fully derived in our present state of ignorance. For example, error probabilities require more input data than specialists possess: It is however entirely possible to develop an error likelihood and consequences analysis, although the quantitative values may not be probabilities, but simple ordinal ratings of error likelihood. It is certainly possible to judge that certain factors induce stress, although the amount of stress (at least in quantitative terms) may be difficult to estimate because, although stress can be induced by outside factors (external to the individual), its manifestation is a function of the individual who cannot overcome those factors, although other individuals can.

All of the subcategories in the TA questions can be estimated, although the precision of the estimates is certainly questionable. TA is not an extremely precise instrument because it is only qualitative; because it is dependent on the specialist's skill, a high degree of reliability among different specialist judgments cannot be expected.

If the system is a simple one, if it obviously makes no excessive demands on the operator/user, then many of the categories in Table 3.2, once reviewed, can be ignored. A TA is important only for complex systems that create complex tasks and serious demands on the operator. TA is important, but only if the system being designed warrants it. However, for complex systems, the analysis, if performed fully, can be prolonged and may create difficulties for the analyst. Designers will not delay design decisions until TA results are available. The extent to which the specialist pursues the TA is a matter of judgment; the specialist should pursue the analysis to the extent that he or she is fairly confident that significant behavioral problems are identified or do not exist.

The reader should note three different elements in Table 3.2: those describing the task (e.g., listing of controls and displays), those involving deductions by the specialist based on the task description (e.g., errors that may possibly occur), and not least the kinds of TA questions the specialist may wish the analysis to answer. These last depend largely on the specialist's skill and experience and his or her deductive logic.

Something should be said about the information sources that may be required to perform a TA. Czaja (1997) listed documentation review, surveys, questionnaires, interviews, observations, and verbal protocols as sources of that information. These information sources have their primary value not in actually performing the TA (i.e., an analytic, deductive process), but in giving the specialist enough information that he or she can begin the TA.

Presumably one analyzes the tasks in new design because they are new. It could be maintained that no task is ever really new and that it is only the new context in which it is performed that makes it new. For example, if a new integrated display must be interpreted, it still requires of the operator certain perceptual/cognitive analyses that may be found in related displays. From that standpoint, learning about a familiar task related to the new one will aid the analysis of the new task. An earlier TA gives the specialist a knowledge base needed to undertake a new TA.

TA is often mandated in military projects. As reported by Luczak (1997), Beevis et al. (1992) indicated that, in a study of 33 military projects, TA played a considerable role. The various uses of TA can encompass the following (taken from Beevis): function flow diagrams (98%) of the projects, mission descriptions (95%), OSDs (80%), information flow analysis (78%), task taxonomy (75%), operator capability (65%), human error analysis (50%), and subjective workload prediction (40%).

One point should be made to explicate the Beevis et al. results. Military projects often demand by name some form of behavioral analysis; in such cases, the military customer is willing to pay for the analytic activity. Under these circumstances, even if the engineering developer is hostile to TA, it will perform the activity.

It is possible that nonmilitary, nongovernmental projects make much less use of task analysis and its derivatives.

Over the years, TA has generated a considerable outpouring of theoretical literature. The following is only a partial listing: Luczak (1997) listed the following besides those already cited in this section: Drury, Paramore, Van Cott, Grey, and Corlett (1987); Diaper (1989); and Meister (1985a). Luczak (1997) also provided references to work published in French and German.

One variation of the traditional TA must be mentioned (with further discussion later). This is cognitive TA (CTA), which is no different in its conceptual orientation from the traditional TA, but concentrates its attention on the decisions required of the user and the preceding cognitive processes. CTA has been employed in development of a number of human performance models including COGNET and MIDAS (see Pew & Mavor, 1998). The task in CTA is decomposed into the stimulus conditions requiring a decision and the factors affecting the user's decisions (e.g., ambiguity of the situation, criticality of the decision, time pressures). The decision, once made, is followed by a response action. (For a more complete treatment, see Hoffman & Woods, 2000.)

DESIGN EVALUATION TOOLS: CHECKLISTS AND SIMULATION

A reminder: Design evaluation involves only the specialist and designer; testing always involves a subject who performs. The common thread among design evaluation tools is that, with the exception of physical simulation, no personnel performance is involved in their use.

Checklists

Checklists are statements of recommended design practice based on research (when the research results are available) and a consensus based on design experience. The type of checklist we are talking about applies only to the HCI; engineers may have checklists relating to other features of the equipment. Checklists are usually verbal, but may include graphic illustrations. Their level of detail may vary from relatively simple prescriptive statements (e.g., "scroll up, not down") to statements with explanations and references to supporting data (e.g., Campbell, 1998). The length of the checklist may vary depending on how much detail one wishes to include. For example, the famous Smith and Mosier (1986) checklist describing computer interfaces includes hundreds of items.

The procedure for using a design checklist is to compare a design output (e.g., blueprint or software scenario) with the checklist and look for discrepancies between the two. This is not as simple as looking for the red marble in a set of blue ones; the applicability of a checklist item to a characteristic of the design output must be ascertained; then the meaning of the item (a concept) requires a translation from the verbal statement to an image of the physical feature in the output. If a discrepancy is found, the designer can then make an appropriate change in the design product.

The checklist has obvious limitations: The principles included cannot be quantified in any way, and there is no suggestion that any one principle is significantly more important than any other (obviously they are, although the specialist cannot assign relative values). This hampers a design change when it is required because the designer does not know how important the change is and how much precedence it has over other considerations. One cannot measure anything with checklists; one cannot add the number of checklist deviations found to evaluate the total product. Each checklist principle is completely independent of every other principle, although checklist items can be grouped according to some overall category like controls, displays, or keyboard factors.

The major function of the checklist is to serve as a memory aid to the designer who can keep in his or her memory only a fraction of all the features of the HCI that may apply. Its use ensures that the specialist has not inadvertently made an error in the design output or has overlooked something.

One of the difficulties in using checklists is that, because they contain only general principles, the checklist may not apply well to an individual design product. This has been a source of continuing criticism — not only of checklists in particular, but also of behavioral design principles in general. If one were to develop on the basis of specific research a set of checklist items that pertained directly to a specific design, then users (i.e., designers and specialists) would be quite laudatory. This is what was done in connection with the automated traveler's information system (Campbell, 1998; Campbell, Carney, & Kantowitz, 1999). However, money and time were made available specifically for the research needed to develop the checklist. One cannot hope to do this with every or indeed most design projects. The Campbell effort demonstrated that a means of overcoming checklist weakness could be developed, but only on a research basis. Obviously no general knowledge resources are adequate to the specific needs of individual designers. A checklist cannot be all things to all designers.

Specialists would love to be able to use the checklist as a design tool; that is, to use it to guide initial design, but this is impossible. All design must be guided by system goals translated into functions and missions.

The checklist is a primitive instrument, but it can be quite useful, and we tend to accept it in design because it is in common use elsewhere (e.g., the procedural checklist used for preflighting an aircraft or energizing a nuclear power plant). Moreover, checklists are in even more common use; the grocery shopping list is a checklist reminding one of what is to be bought.

Simulations

A simulation attempts to reproduce the system and the conditions of its use. It does so in various ways, ranging from physical reproductions to entirely symbolic ones. As a design evaluation tool, the simulation attempts to exercise a preliminary or final design product by putting it into a use framework.

There are four varieties of simulation:

1. *Physical simulations.* These reproduce in full scale the system under development and permit its exercise under conditions that reproduce the operational environment (OE). Examples of such simulators are those that represent aircraft, ship handling, sonar detection systems, and automobiles. These simulations range from extremely faithful complete reproductions to part-task (partial) simulators.

2. *Behavioral simulations.* The most common variety of these is called the *walkthrough.* The walkthrough makes use at an early design stage of paper reproductions (e.g., blueprints or scenarios) of the system or system components. As we see later, walkthroughs can be performed with operational equipment; these are used for knowledge elicitation and gathering data for a cognitive task analysis (CTA).

3. *Symbolic simulations.* These are human performance models that are computerized and describe the system and its operations symbolically and dynamically. Procedures for running the system model permit one to

exercise that model in a mission mode (from initiating operations through accomplishing the mission goal).

4. *Computer simulations.* These present the system and/or its modules like the HCI in graphic form on the computer display. Such computer simulations are used primarily (although not exclusively) to display the HCI and the stimuli it presents to an operator. The most sophisticated use of computer simulation is virtual reality.

A discussion of simulation in its varied forms would require a textbook of its own. We confine ourselves to simulation as applied to design processes. Readers interested in a more complete treatment of the topic should consult Laughery and Corker (1997).

Each of these simulation modes enables the specialist to examine a number of design questions (e.g., which of several alternative configurations is most likely to be effective; what errors are likely to be made by an operator; whether a particular mission procedure can be performed; whether system personnel can perform their tasks adequately to accomplish system goals).

The various simulation modes differ on a number of dimensions:

1. Fidelity to the physical system and operational environment (OE). Aircraft simulators may reproduce an aircraft cockpit completely by using actual controls and displays in an operational configuration and by using motion simulation and optics to reproduce the physical and visual sensations of flight, including emergencies. Fidelity can be reduced from almost perfect reproduction of an operational system by using substitutes for operational equipment.

2. The nature of the mission can be varied. For example, part-task simulators have been developed that reproduce only segments of the mission.

3. Whether a human test subject is involved in the simulation. Physical and computerized display simulations always involve at least one human; symbolic simulations (human performance models) do not involve a human, although the effects of a simulated human (represented by distributions of possible responses or deterministic equations) are incorporated into the model.

4. Cost constraints and time obviously influence the various simulation types. Behavioral simulations cost almost nothing; physical and symbolic simulations are very costly.

5. Another factor to be considered is the effectiveness of each simulation type for its design use.

Some simulation types are more useful for certain design questions than are others. Physical simulations are obviously quite effective for evaluating performance because they reproduce the system and its environment almost completely. One cannot test actual personnel performance in a symbolic simulation, although one can use the model to predict that performance. One can more adequately test differences between configurations in a physical simulation than in a symbolic one, although theoretically the human performance model could be used for this purpose.

Because the human performance model is presently most often used only for research purposes, it cannot be said that symbolic simulation is effective for actual design. For its very limited purposes, the behavioral simulation is very effective.

Let us consider each simulation type in somewhat more detail.

Physical Simulation. Everyone is familiar, if only by reputation, with elaborate, highly sophisticated simulators of the aircraft type. Simulators of this type can reproduce the system, its physical and operating characteristics, an entire mission, with great precision. Every new aircraft system has a simulator built in parallel with ongoing design. This facilitates performance measurement of proposed new modules. For example, one might plug various navigation subsystems into the simulator and test their comparative performance. These simulators are often designed so that they can be physically reconfigured in various ways, allowing design variations to be tested.

Automatic measurement is often one capability of the simulator. This feature is necessary for testing of complex systems because so many events occur simultaneously that one cannot rely on even a corps of measurement personnel to collect the data. Automatic data recording permits measurement at an extremely detailed level, which may be necessary to answer design questions.

The simulation is an ideal vehicle for operational system testing where, as in nuclear power plant operations, it is not feasible or desirable to power up the actual system. The downside of the physical simulator is its cost, which restricts its use to a small number of projects usually supported by government or major industries.

The measurement use of a highly sophisticated physical simulator is much like a laboratory experiment. One does not casually enter a simulator and say to the technician in charge, "plug this module in," and proceed immediately to measurement. Considerable test planning is required.

Behavioral Simulation. This is like the first reading by the cast of a new play entering rehearsal; only the words are reviewed (in our case, only the procedures involved in system operations are reviewed). The behavioral simulation is a walkthrough, or talkthrough, or, with cognitive tasks, thinking out loud; the cast walks through its lines. For example, the drawings of the HCI, which have been developed to a certain stage, are reviewed by the design team; as each task is read, the design team asks questions (e.g., Will the display be fully visible to the operator? If two almost concurrent actions must be taken by the operator, will he or she have enough time to perform the actions sequentially?). This is a fairly molecular examination, task by task. The procedure is analyzed as it is reviewed to discover any behavioral problems that may have been concealed, so that in one sense this is an evaluative or verification test as well as an analysis.

The walkthrough is most effective when the mission is largely procedural; if the operator's actions are covert, as in monitoring a display, not much can be done with these in a walkthrough.

Now that computers are available, the HCI may be presented on the computer terminal screen, but otherwise the process is as it was described previously. Nothing is done in real time; the walkthrough is essentially an analysis stimulated by the task reading and any visuals presented. The talkthrough is essentially the same thing depending on how much the verbal protocol is emphasized.

The great advantage of the behavioral simulation is that it costs nothing except time and attention, requires no elaborate instrumentation except possibly a small amount of computer programming, and can be performed as early as the first function/task description and the availability of the first rough equipment sketches, scenarios, or storyboards.

The uses to which a walkthrough can be put include: (a) review of the adequacy of the procedural steps in the mission and their interrelationships; (b) suggesting where procedural errors are likely; (c) indicating where special knowledges and skills may be required; and (d) comparing alternative procedures and/or design configurations.

Symbolic Simulations.   The wave of the future, if one can believe the theorists, is the human performance model, which depends, of course, on computer capability. This section is not going to be a complete (or indeed even partial) treatment of the topic described at length by Pew and Mavor (1998).

The human performance model is a symbolic representation of the system (in whole or in part) as it is operated by personnel. Ideally the model describes the physical components in action, the environment in which the system functions, and the tasks performed by personnel. The model is essentially inert until it is energized by the input of data describing model elements (e.g., in the case of personnel, the distribution of human response times and error probabilities). The model performs in accordance with a predetermined set of rules, such as the task with the higher priority (entered as part of the input data) is performed before tasks of a lesser priority. When energized, the model progresses from one mission phase to another on the basis of those rules.

The outputs of the model exercise are a set of quantitative values (as specified by the creators of the model) that presumably predict whether system goals will be achieved, the length of time it will take to perform the mission, and so on. The model output is therefore a prediction of the performance to be expected with a specific system configuration. The model can be exercised any number of times and, once input data are inserted, it requires no further human intervention. The number of times a model is exercised depends on how much confidence the user wishes to have in the model outputs.

Most models are stochastic, which means that they provide only probabilistic values, which of course vary somewhat over time with each exercise of the model. Only a few models are deterministic (i.e., output values depend on the solution of equations).

The human performance model can be used to compare alternative design configurations and suggest how well personnel perform with those configurations. Because it is run on a computer using a task network, produced by a general-purpose engine like SAINT, it is relatively easy to introduce behavioral or equipment modifications by changing input data or rewriting input code.

Model outputs can provide useful information to the design team if the model can be energized early enough in the design process (it is absolutely of no value at the end of design). Thus, it anticipates final operational testing (OST), but cannot replace the test because the human actions in the model are based not on actual people but on statistical distributions of how people are presumed to perform. One of the positive features of the model is that it can be exercised repeatedly, thus focusing and sharpening its predictions. Although actual testing is more realistic, ordinarily the test is performed only once, which means that there is somewhat less confidence in the predictions it supplies.

The great advantage of the model is that it can be exercised as soon as some equipment details are available, and of course it can be run as many times as desired. The model's disadvantage is that it requires considerable equipment and human input data. The former can be secured from engineering; the latter depends on prior research, which may or may not be available.

Human performance models have been around since the early 1960s, if not earlier. They are, however, largely experimental, funded primarily by governmental agencies, and it is not clear to what extent they are utilized routinely in system development (see Chaffin, 1997). Pew and Mavor (1998) and Davies (1998) reviewed the status of such models and described some of the problems involved in using them. Davies supplied a British viewpoint.

There are various types of models, which Davies classified as follows: control models, sensory models, anthropometric models, workload mod-

els, human error models, and task network models. The following considerations, phrased as questions to be answered, apply to the selection and
use of a model:

What is your specific goal/problem domain?
Certain models have been optimized for specific problem domains and
are preferred to other models (e.g., if the specific problem area is associated with control design the optimum control model [OCM] would be
most appropriate).

At which stage of the design life cycle do you wish to apply the model?
Some models are more suitable for early conceptual design, whereas
others can be used during test and certification phases.

How much time and man hours do you have available for the assessment?
Time budgets are important particularly in a fixed price environment.
Some models vary widely in the time required to learn how to use
them, provide them with data, and so on.

*What type of human resources do you have available (e.g., HF specialists, sys-
tem analysts, software specialists)?*
This is an important discriminator among the models because some of
the models require extensive knowledge about HF to use correctly,
whereas others require the assistance of software experts.

How much hardware and software budget do you have available?
The costs in procuring the necessary hardware and software vary
across the range of available models, and budgets need to be taken into
account.

Which operating systems do you have available?
One's hardware budget is important here also; if one only has a simple
PC-based system and wishes to carry out a sophisticated mission anal-
ysis, for example, the PC may not be adequate.

What model capability do you require?
This question determines the processing requirements of the model; for
instance, if it is necessary to incorporate a cockpit layout, then models
like TAWL/TOSS will not be suitable, but models like MIDAS or HOS
will be fit for the purpose.

What type of data are available as inputs to the model?
Some models require a task analysis, whereas others may require detailed timing information. If the system designer has a task analysis in a
timeline format available, then it would be inappropriate to recommend a model that required a task network format.

*What type of output do you need (subjective ratings, reaction times, error
rates, timelines)? How much precision do you require?*

The outputs of models vary, and the designer must examine the possibilities with regard to their outputs.

Davies noted certain shortcomings in her review of such models:

• Theories underlying the models were often either unknown or in some cases fallacious; few models were fully validated or verified; there was difficulty in validating complex models; tightly coupled, interdependent components allowed systems to perform more effectively, but limited development of new components and sharing/receiving of modules with other organizations;

• Models had to build in sufficiently general procedures (so coding could be re-used);

• Models had difficulty in incorporating global effects like fatigue, age, training level;

• There was difficulty in quantifying the value provided by using the model as part of the design process;

• Models lacked acceptance criteria for workload measurement and had difficulty relating performance to mission effectiveness;

• Support for models was often dependent on company funding;

• There was also some difficulty in assessing the quality of the interface in terms other than workload.

Hence, human performance modeling is something to be pursued more effectively in the future. If validated, the models could be of great benefit to designers. However, it seems unlikely that the models are used extensively in present engineering development, although the U.S. government (DOD) intends to pursue research on them. This was indicated by a recent article by a high official, Hollis (1998), who used the term *simulation-based acquisition* to include and be heavily based on human performance models.

Computer Simulations.   It is now possible to use the computer terminal screen as a means to present stimuli that replicate elements of the physical system. The tremendous explosion of sophisticated computer graphics permits this. For example, the screen can portray a work station with controls and displays; the latter can present changes in digital and scalar values to represent changes in system status. Using a keyboard or mouse, one can modify these displays, call up system status information dynamically, and so on. It is therefore possible for a test subject to interact with a simulated work station in much the same way that he or she would if faced with the actual equipment. In fact, the simulated work station, when optimized, can become the actual work station for the PC. Obviously there is a difference between a physical instrument and a representation of that instrument in the form of an outline figure on a screen, but this difference, if tested, would probably prove insignificant.

The computer simulation can be very useful: to compare various interface configurations and test human performance with a single configuration. For example, anthropometric models can be used to evaluate visual sighting angles. The computer simulation can be used as early as initial design analysis and as late as detailed design.

PROTOTYPING

Prototyping has been repeatedly mentioned in these pages because it is a tremendously popular design methodology (see Hartson & Hix, 1989; Shneiderman, 1998). Essentially, it is the process of developing versions of a design product, either fully or in some abbreviated form, and presenting these to a group of potential users for their opinions, preference, or even performance. Subject responses are then fed back, for further design consideration, into a more refined version, which can then be presented to a new user group who then provide their feedback, and so on.

The advent of the computer made it easier to develop the prototype designs and then present them. Before the computer, it was necessary to build a physical representation (a mockup) in wood, cardboard, styrofoam, and so on, and this made it impractical for other than the solution of critical design problems. Outside the computer simulation, there is nothing tremendously novel about the process; it involves performance measurement, opinions, and preferences collected using traditional psychological test procedures.

As in any performance measurement situation, questions arise and must be answered: (a) selection of the subject population, who they are, where one finds them, and how many should be used; (b) what stimuli should be presented (this is the prototype design; how much of the total system should be presented); (c) where the performance tests should be held (at the engineering facility, in some operational area, elsewhere); (d) what kind of responses (measures) should be required of test subjects; (e) what kind of measurement instrumentation is required (paper-and-pencil scales, tape recorders, timers, cameras); (f) how the data are analyzed (this decision should of course be made before selecting measures); and (g) how many times the prototype test is repeated. This is all familiar to students of psychological research.

One of the attractions of prototyping is that it can be accomplished relatively rapidly (hence the term *rapid prototyping*). It can be performed quite informally (e.g., using coworkers as subjects, collecting data on the engineering floor, etc.). It can even be performed without formal planning, although questions have been raised about the rapidity and informality.

To a certain extent, prototyping permits the implementation of user- or human-centered design theory. The more one uses the procedure, the greater user influence there is on the development of the system. However, that is hardly the reason for the great popularity of the method. Prototyping is a relatively simple way of making design decisions. Rapid prototyping can also be viewed as a means of shortening the design process (i.e., avoiding extensive analyses) and thus saving money. If questions are raised about design adequacy, the design team can always justify itself by referring back to the results of the prototype tests.

It is not clear whether there have been empirical comparisons of prototype design effectiveness as compared with designs not using rapid prototyping, and it would be difficult to make this comparison because it would be necessary to conduct two design processes independently (two sets of designers with their idiosyncrasies) and in parallel. The enthusiasm with which the methodology has been adopted suggests, however, that it has a certain value for the designer, which a negative appraisal would not be likely to overcome.

COGNITIVE DESIGN TECHNIQUES: COGNITIVE TASK ANALYSIS AND ECOLOGICAL INTERFACE DESIGN

Cognitive Task Analysis

There has been increasing interest among HF professionals in cognitive functions and their relationship to design. There is even a buzzword, *cognitive engineering*, that encapsulates that interest.

Paradoxically, the concern for the human (and hence for his or her cognition) has increased because of the rise of the computer. If the mundane work of controlling machines has been given over to the computer, what else is left except cognition? If there is an operator, what else is there for this operator to do except perceive and think?

This section discusses cognitive task analysis (CTA) and ecological interface design (EID) as they relate to cognitive engineering (design). Therefore, we might begin by trying to define what cognitive engineering actually is. There is considerable confusion about this term and CTA.

Cognitive engineering is system design predicated on the user's expertise and knowledge (McNeese, 1998). Czaja (1997a), basing her definition on Woods and Roth (1988), reported that cognitive engineering is an applied cognitive science that relates the knowledge of cognitive psychology and other related disciplines to the design of cognitive environments (e.g., decision support systems; see also Woods, Watts, Graham, Kidwell, & Smith, 1996).

What cognitive engineering suggests is that systems have become so dependent on cognitive functioning that their design depends on determining and applying what system users think. There is a link here with prototyping, which is an effort to discover how potential users view new design. Indeed, Potter, Roth, Woods, and Elm (1998) talked of using prototypes as tools for discovery of how users think.

To discover what and how humans think, it is necessary to utilize CTA. What kind of design problem is representative of cognitive design for which CTA would be most useful?

Imagine a requirement from the CIA to develop a computerized system that reproduces the analyses performed by intelligence agents who collate and interpret information from a variety of sources. This task has been performed for many years, but no record or analysis has been made of the way in which they performed the job. This is where CTA enters because the system designer must develop an exhaustive atlas of the agents' thinking processes. Potter et al. (1998) cited other examples of systems to which CTA has been applied: aeromedical evacuation planning, command and control operations, commercial aviation, anesthesia, and nuclear power plant emergencies.

CTA is not a single method; rather it is a basket of several methods that the reader may find quite familiar: critical incident reporting, knowledge elicitation (used mostly to develop expert systems, but a cognitively designed system is a form of expert system), direct observation of what the user does in the field, walkthroughs and talkthroughs, simulations, and cognitive mapping. These are methods to secure information that is then decomposed into categories like (a) stimuli, (b) effective cues, (c) type of data interpretations made, (d) effect of imprecise or incomplete data, (e) decisions required, and (f) factors impacting these decisions. The judgment categories are determined by the CTA analyst alone (there are no standardized categories) and may be displayed in verbal form (e.g., as a set of vertical columns) or graphic form (e.g., arcs representing relationships among task elements).

Having developed the CTA, the next and somewhat daunting problem is how to apply CTA results to actual design. This is the classic problem of transforming behavioral principles and data into physical structures: Cognitive engineering is not clear about how this is done, but there is an implied assumption that detailed analysis of the design problem in cognitive terms automatically leads to a more effective design. Maybe so.

CTA is probably significantly more difficult than traditional task analysis because cognitive functions are much more complex than procedural ones. CTA is therefore still a somewhat experimental methodology particularly because the collection and analysis of data are very time-consuming.

Ecological Approaches to Design

Computerized systems focus on displayed information. As a consequence, there has been increasing interest in display design. Cognitive specialists have taken a number of approaches to display design. Flach, Vicente, Tanabe, Monta, and Rasmussen (1998), from which much of the following discussion has been taken, have differentiated four types of design approaches, which they characterized as technology-centered, user-centered, control-centered, and ecology-centered. In this section, we emphasize the ecological approach because it appears to be the most theoretically sophisticated and may encompass the others. Before considering it, however, we propose to review more traditional approaches to design.

Technology-Centered Approach.   This approach centers on the capabilities and limitations of technologies. What can the technology do? How high, fast, and far can the new vehicle fly? The emphasis of the technological approach is functionality. The system interface is typically designed in a way that reflects technological capabilities (e.g., a display for every sensor and a control for every control surface). This can lead to a proliferation of displays (in aviation, warning indicators, status displays, air traffic control data links, meteorological information, communications data, etc.). With each new innovation in sensor or automation technology, a new display is added; this expansion of data presentations has on occasion exceeded the information-processing capacities of the operator.

The User-Centered Approach.   This has tended to focus on the limitations and capabilities of human operators and the implication of these limitations for how systems should be designed. This approach has emphasized these questions: How much will operators have to perceive, do, remember? What problems will they have to solve? How do design features influence the kinds and probabilities of human error? The principal objective of this approach is to ensure that the demands of operating a new technology do not exceed human thresholds (e.g., limited information-processing capabilities) of the operator.

From this perspective, the capability and expectations of users are important concerns for designing the system interface. The presentation of information should conform where possible to user expectations (e.g., population stereotypes or mental models), and the information should be integrated into a small number of chunks so as not to exceed human working memory capacity. Principles such as stimulus–response compatibility, and stimulus/central processing/response compatibility are involved (Wickens & Carswell, 1995).

The Control-Centered Approach. This focuses on the coupling between humans and technologies. The human is viewed as a controller or supervisor of the technology, which is generally characterized in terms of its dynamics. These dynamics can be represented by a set of first-order differential equations — state equations. The variables in these equations are called *state variables* or *system states,* and the coefficients represent the action constraints that determine the mapping from current to future states.

The particular concern of this approach is on the stability of the human–machine control loops. This involves the order of control (i.e., the number of state variables needed to characterize the dynamics) and the time delays (i.e., the time it takes to respond to an input). High orders of control and/or long time delays can result in instabilities (e.g., pilot-induced oscillation). From the control perspective, a fundamental concern is observability or transparency — all state variables are measured and represented in the display and time delays, and noise associated with the measurement and display process is minimal. One strategy for reducing the delays associated with information processing is to support the operator's capacity to anticipate future states through predictive displays, quickening, and compensatory formats.

Another important concern for the control-theoretic approach is how the system is controlled. This is typically reflected in the organization of components (human and automated) within the control system. An important consideration is whether the human is required to function as a manual controller (*hands on,* as it were) or as a supervisory controller of an automated system. Automation can often increase local stability by reducing the time constants in the loop (e.g., as with autopilots). However, automated systems have been characterized as *brittle* because they function well only as long as certain assumptions about the control context are satisfied (e.g., assumptions dealing with linearity or stationary dynamics). When these assumptions are violated (e.g., the dynamics of an aircraft can change at high altitudes), the system becomes unstable without aid from a supervisor. Humans, however, have difficulty keeping up with the demands of the system, but tend to be more robust in creatively responding to unexpected events. The fundamental question is, how can one link human and automated systems together in a way that satisfies constraints on stability?

The Ecological Approach. The previous three approaches differ in terms of which constraints are emphasized: the capabilities of the machine (the technology-centered approach); the limitations of human information-processing systems—for example, workload (the user-centered approach); and the stability constraints arising from coupling between the human controller and machine dynamics (the control-centered approach). Despite these differences in emphasis, the three perspectives have a common image of the system: The system consists of the human and machine alone, apart from its problem-solving situation, its context, and its environment, all of which EID emphasizes.

The ecological approach starts with a broader view of the system (Vicente, 1999). This includes the work domain (work space or problem space) as an integral component of the cognitive system. The first three approaches define the system relative to its structures (i.e., human and machine), and the ecological approach defines the system relative to its function within a larger work or problem space (i.e., its ecology). Considerable emphasis is placed on work context; this is a determining characteristic of all the nontraditional approaches to design that we have encountered so far. In information terms, the human is viewed as a *meaning* processing system, rather than as a transmission channel with limited capability. Technology is viewed as an instrument or a tool that is best appreciated in terms of its function relative to a work space (ecology). The human may often interact with the work domain through a computer or automatic control system, but it is the interaction that is the ultimate domain. The question to be answered in design then becomes not only the interaction between humans and machine, but also and rather the interaction between humans and the total work domain. The constraints that matter in terms of work become fundamental; constraints on information processing, technology, or control loops become secondary. This does not mean that information processing, design technology, or control loops become irrelevant, but rather that these constraints can best be appreciated relative to the total situation or context.

The term *ecological* was adopted to emphasize the focus on constraints arising from the work ecology—a focus shared by Brunswik's (1956) and Gibson's (1979) approaches to human performance. The term *ecology* is used instead of environment to emphasize the reciprocal dynamic between the operator and his or her work niche. The term *emphasis* is used to indicate the relative weighting of the approach, rather than a rejection of the other approaches.

Ecological theorists insist that questions must be framed in the context of something called a *dynamic distributed cognitive ecology*. What this means perhaps is that the need for understanding is not restricted to the single display, but extends to a variety of performance shaping factors. For example, it is not enough for pilots to know their current altitude. They must also know the significance of that altitude relative to their flight objectives (e.g., landing, evading detection by enemy radar, etc.) and relative to the dynamic capabilities of the aircraft. (Note a relationship to another concept, *situation awareness*; Endsley, 1995.) The concept is that operators are actively seeking information; thus, the cognitive performance shaping factors are integral components within the dynamic system. They are not just additional sources of noise and time delay. The questions that the human operators ask are shaped by the situation (i.e., the context or work domain; in connection with work domain analysis, see Miller & Vicente, 1998).

There are multiple layers to the question of the correct approach to behavioral design. These other more traditional layers/approaches do provide valid insights into the phenomena of human work; ecological theorists see the approach one selects as simply a matter of depth of analysis. Questions about how information should be displayed cannot be adequately addressed within the narrower scopes defined by the three traditional approaches. Ecological theorists believe that their approach represents them all but adds an increment of meaning that the others do not have.

Design Practice

This chapter attempts to portray what human factors (HF) specialists working in system development actually do. Such descriptions have been written before (Meister, 1971, 1991), but have generally been one author's perception of what goes on. What the authors have done is to ask a reasonably large sample of people what their perception of the design/development process was. We are dealing with perceptions here because the process is so complex, functioning over masses of time, and personal experiences are so variable that all one can extract from respondents are perceptions only. However, if there is sufficient commonality in respondent replies, we can say that these perceptions are veridical because what is perceived by a sufficient number of people becomes the truth about what they perceive, although not necessarily an objective truth.

To study the design/development process, we developed a questionnaire and sent it to a group of people who were involved in design/development. The initial survey group was composed of members of the System Development and Computer Sciences technical interest groups of HFES. To supplement these people, we selected members of the society who were employed by major industries: Boeing, IBM, and so on.

The survey described next was organized to elicit judgments of how often and/or important a particular practice was performed by respondents. The questions posed represented what we consider to be the central aspects of HF work in development. In all, there were 53 statements of this type: "Task analysis is _____ employed from the start of the design process," in response to which the respondent was asked to check one of five responses: *always, usually, sometimes, rarely,* or *never.* In other words, a

87

five-point Likert scale was used. Other items asked the respondent to indicate the importance of a particular method. We were particularly interested in determining the cognitive strategies used by HF specialists.

We are aware of the potential difficulties of a survey. This is an intensely subjective methodology. Apart from the subjective aspect of a questionnaire, there are other problems, such as the vagueness of the terminology employed in design and development. For example, the terms *design* and *development* are often used in tandem. Are they the same? If one is used in place of the other, will a majority of respondents reply as if they are the same or different? If different, to which one are they referring when they are asked a question?

The best way to study design practices is to (a) observe what the designers and behavioral specialists do in the course of designing an actual system, equipment, or product (SEP); (b) ask them to keep a diary of their activities and thoughts; and (c) interview them directly. None of these can be done without extensive funding, which the authors did not have. Later we describe several studies for which funding was provided.

We went with our best shot, knowing in advance that any questionnaire study of design practices must be less than completely adequate, but reasoning that, despite these inadequacies, a questionnaire might be able to dredge up useful information in an area of HF about which too little is known.

Is the topic of how designers and specialists function in design important enough to warrant extensive research? We believe the answer is yes for the following reasons:

1. Our technological civilization depends on how well engineers and HF specialists do their job. From that standpoint, the description of HF in design is a contribution to the cultural history of the 20th century. More practically, until we learn more about how design personnel do their jobs, we cannot know whether improvements are warranted. We also cannot derive the needed information merely by examining the final design output because this merely represents the culmination of a long, complex process.

2. Arguably, HF involvement in the design process is the one feature that makes HF unique — the marriage of the physical and the behavioral (see Meister, 1999, on this point). Consequently, it is important to learn more about this aspect of HF.

3. The fact that design is the solution of a problem indicates that, although engineers deal with physical problems, forces, and objects, design is at bottom a behavioral, cognitive process. From that standpoint, it deserves as much HF attention as does research on aging, perception, or computer software.

THE SURVEY

The questionnaire was organized around the following points:

1. Demographic information: name, years of experience, highest degree, type of system worked with ordinarily.

2. Frequency and utility of task analysis (TA).

3. Elements of the customer specification and their importance in subsequent phases of the design/development process.

4. Importance and frequency of use of such common methods as TA generally and more specifically its derivatives (e.g., operational sequence diagrams [OSD] and flow diagrams).

5. Logic and rationality of the development process. This question was asked because there are contrasting views of the process (see Chap. 2): that it is essentially rational or essentially chaotic.

6. User participation in design. This aspect has been assuming increased importance, at least as described in recent research reports. The question is the degree of such participation.

7. Respondent's experience with initial design, update, or redesign. Most descriptions of the design/development process have assumed that most design is initial, but there is reason to believe this is not completely the case. In any event, the activities involved in each may be different.

8. Nature of test and evaluation (T&E) and particularly the formality of the testing. T&E descriptions usually posit formal procedures, but is this actually the case in the usual development process?

9. Whether HF specialists get involved in developing operating procedures. The link with TA is obvious.

10. Theorists emphasize the importance and utility of system model simulation as an integral part of design. Utility and importance are accepted, but the question is whether system simulations are actually employed or is this essentially a theoretical concept pushed by government research and development (R&D), but not used much by industry. There is the associated question of whether such simulation actually proves useful in design.

11. A number of questions dealt with the specialist's relationship with the primary designer. One question asked whether the specialist is asked to justify any conclusions he or she may present to the designer.

12. A central question in this book is how useful the HF literature is to specialists and designers. Previous work has questioned that usefulness. We sought some clarification on this point.

13. Descriptions of the design process have suggested that the comparison of the design alternatives can or should be a formal, quantitative

process (Meister, 1985a). The question is whether such a formal process is followed or whether it is entirely informal.

14. What is the strategy (if indeed there is one) that the specialist pursues in analyzing/evaluating interface designs? Does the specialist have a strategy or does he use memorized checklist items or broad generalities like KISS (Keep it simple, stupid!)? (In this connection, see Lund, 1997.)

15. In the past, specialists have sought for authority to reject interface designs they considered poor. Few specialists had such authority. Do they have it now?

16. The question of whether operator requirements are included in a customer specification is extremely important to the specialist's work. Physical requirements provide direction and force to equipment designers. Absent such requirements for operators, the force behind HF work in development is much diminished. In the past, such operator requirements were not included in the customer specification. Has the situation changed any?

17. An aspect of the strategy to be followed in analyzing the customer's requirements is to try to imagine the operator functions to be performed by the total system. To what extent is this strategy followed?

18. One theoretical question that has absorbed writers on design/development is whether this is a top–down, bottom–up, or combined process. How one conceptualizes the overall process is quite important in directing any research on the topic.

19. A general question is whether there is a perception by specialists that HF is adequately considered during development. In the past, we had the impression that most specialists did not feel that it was adequate. Is this impression still true?

20. The specialist has available to him or her certain resource documents, previous test reports, or technical manuals from a predecessor system. Are these really useful?

21. Descriptions of design/development represent the process as iterative (i.e., with feedback affecting prior decisions). Is this true?

22. If–then reasoning (propositional logic) is assumed to be a major part of the design process. This is a logical approach rather than a highly creative, intuitive one. Given that a certain human function is required by the system, certain consequences follow; these determine lower level, more molecular design. The question is: To what extent do designers and specialists engage in if–then reasoning?

23. To what extent do specialists refer to design guidelines and standards, like Smith and Mosier (1986) or MIL-STD 1472F (1999)?

24. TA or an operating procedure is assumed to be critical to the design of the HMI. To what extent is this true?

25. The nature of the relationship between specialists and equipment engineers has been a topic of some concern from the beginning of the discipline. An element in these relationships is the extent to which designers are willing to accept HF recommendations. Two items dealt with this acceptance.

26. One of the constraints generally bemoaned by writers describing HF design processes is the insufficiency of time allowed for analysis and evaluation of outputs. It is generally assumed that system development is a frenetic process. Is this true?

27. The availability of prior system documentation is also important considering that much of design builds on prior system design. Does the specialist find this information helpful?

28. The question of users was asked again to ascertain the respondents' reliability. How much is the user consulted during design?

29. Certain problems may arise during development for which answers cannot be supplied from previous research or the best guess of subject matter experts (SMEs). The preferred solution would be to perform research to answer the unanswered questions. In the past, we assumed that there was neither time nor money to perform this research. Is this still the case?

30. A common objection of specialists to the way in which design is performed is that they are frozen out of that design until after major decisions are made, thus limiting their usefulness to post facto critiques. Has this situation changed any over the years?

31. A related question involved the specialist's participation in major design decisions. Because these are made early in development, the specialist's involvement in the process is quite important. Once these decisions are made, everything else is derived from them. One question asked whether HF participation in design had increased.

32. A great deal of emphasis is placed on the function-allocation (FA) process. Our point of view is that FA derives from the essential nature of the design process. In any event, considering the importance of FA, to what extent does the specialist contribute?

33. The same reasoning applies to trade-offs in design. Considering the complexity of design with multiple interests like cost, reliability, and so on competing against each other, we are naturally concerned with the question of whether the specialist is involved in such trade-offs.

34. Are standards used and to what extent?

35. How formal is the FA process? Theorists from Fitts (1951) on have described it as a process almost independent of design. Our point of view is that it is part of the total system design in context and that, rather than being independent, it is a derivation of design. One develops what ap-

pears to be an adequate design; only then does one consider the human functions involved; and only if these negate a particular design option does one pay much attention to FA. Is this true?

36. Rapid prototyping is a common software testing procedure. Earlier we discussed the advantages and disadvantages of this combined design/ T&E procedure. One question dealt with the frequency with which rapid prototyping actually is performed.

37. How useful are the various behavioral design methods like TA and OSD in design? Theorists write as if these methods are essential to design, but are they? One possibility is that design decisions are made first and these methods are used to provide documentary proof of the correctness of these decisions. This is a somewhat cynical point of view, but is it correct?

38. It is assumed that design is a team affair. Is the specialist usually part of such a team?

39. Another point that theorists write about is the formality of the design process. In part, any description of design is bound to read as a formal process because use of language forces this. We suggest that there may be much more informality. One item dealt with this point.

40. What are the essential differences between hardware and software design and between military and commercial design? Software design is strongly oriented to cognitive, flexible functions with more options for the operator; hardware design is more proceduralized, with fewer options. Military design caters to a single customer (usually, not always) and commercial design to multiple customers. In software and commercial design, more attention must be paid to user desires. Two questions sought to determine whether these presumed differences actually cause a change in designer strategy.

41. Many years ago, a communications specialist (Allen, 1977) indicated that when designers needed an answer to a question, they often used fellow designers to give them the answer. This is of course the easy, lazy way of using resources. To what extent is this true?

42. Our design efforts start with functionality, which means that they begin with and are most concerned about operation. However, every equipment ultimately fails. Design for maintenance is particularly taxing because it deals with symptomatology analysis and diagnosis, which are inherently more difficult than operation and monitoring. To what extent do specialists design for maintenance?

43. One can look at design as a highly creative, intuitive process (although see later discussion of this point) or as highly rational based on consciously deductive (if–then) thinking. We do not believe that intuition necessarily involves an emotional state; it may be based on successful ex-

perience. We hypothesize that the experience may have been forgotten, but the strategy for the experience is retained in memory. One item sought to determine the extent to which intuition enters into the design process.

44. Ideally, a formal quantification procedure should be followed in the selection of the one best design alternative. Do specialists or designers utilize such a procedure?

45. Computer-assisted design (CAD) is accepted as a part of most design efforts. Does the computer do the thinking for the designer or does it merely build on prior human thinking processes? The thrust of this question is the extent to which the designer is still critical in design, with the computer being merely an aid. Or has the computer become dominant?

46. Organizational theorists emphasize the importance of sociotechnical factors in design. To what extent does the equipment designer or specialist, working at the work station design level, make use of sociotechnical concepts?

Whatever the responses to the questionnaire (see next section), these are the questions that need to be explored empirically. A theorist can come up with what appears to be logical, reasonable answers to these questions, but how do the answers relate to real-life development?

RESULTS

General

What we asked our respondents to do was describe their perceptions of the design/development process. A perception may or may not faithfully reflect the true state of affairs. The situation is further complicated by the fact that we ask, in some cases, about an objective situation and, in other cases, about how the individual feels about that situation. To counter this uncertainty, one can make a heuristic assumption that the existence of a perception makes that perception at least partially true because design is largely the consequences of the activities of its personnel, and those activities are, in part, driven by their perceptions. Therefore, the perceptions reflect reality.

In any situation in which one endeavors to describe a group perception, a difficulty arises when responses to a question are fractionated with no clear unanimity.

Two kinds of results are possible:

1. A preponderance (e.g., 85%) responds in the same way, and the responses (although diverging in frequency) are clustered in the same

half of the response continuum (e.g., 20% *always*, 30% *frequently*, 35% *sometimes*).

2. There is some marked disparity in the responses, so that we have one minority opinion (e.g., 40% *frequently*, 30% *sometimes*, and another minority at the other end of the continuum, 30% *rarely*).

In Situation 1, can one treat these responses as valid reflections of a true situation? Probably not completely, but the consensus gives one more confidence in the validity of the responses.

In Situation 2, it is necessary to explain how the disparity came about. Obviously respondents to this question have different perceptions of what is presumably the same situation. (Or perhaps the situation is not really the same for all respondents.)

Yet those who provided disparate responses to one question also provided consensual (homogenous) responses to another question. Does this mean that one can believe the consensual responses but not the heterogenous ones? This is not logical. We would rather believe that all responses are meaningful, but that the situation described in the question is influenced by factors that produce divergent responses. These factors might be:

1. Homogenous responses are activated by a strong belief system. For example, there is great agreement that users are contacted during design. That is perhaps because it is intellectually correct to believe that this is what should occur, given the great emphasis on the desirability of user-centered designs.

2. The individual factors that affect design/development influence different specialists in different ways. One example is the nature of the system one works on: Design of a large-scale, sophisticated system may require one kind of activity of the specialist; design of a relatively lower level (e.g., work station) system may require a different kind of activity of the specialist. We know from other sources (e.g., Meister, 1999) that the specialist's work is not monolithic. From this standpoint, all perceptions may be valid because they reflect genuinely different situations.

3. Specialists may be responding to different aspects of the same situation. For example, if we ask whether hardware and software design/development are like each other or very different, we encounter a situation in which the overall developmental process (represented by the higher order phases of analysis, design, and test) is the same for both situations. However, the individual design activities within the larger developmental context are very different. Respondents who said hardware/software design was the same may have thought of the higher order developmental

context; those who said hardware/software design was very different may have been thinking of very specific (more molecular) activities.

Another possibility is that what can be called *local conditions* influence the design practices of the individual professional. In Condition A, a particular software company has its own style manual for software and insists on its personnel using it rigidly. In Condition B, another company allows its people much greater freedom to use general industry guidelines. Each situation might influence the design process in different ways.

Although we would never expect design practices to be monolithic, the wide variation in responses strongly suggests that some performance-shaping factors (about which we know little) are involved. It was also considered possible that some of the variance observed resulted from differences in the working environment of those primarily engaged in software design as opposed to those working with less specialized systems (involving both hardware and software). Professionals working more intensively in software development might have somewhat different responses to the survey.

A secondary analysis was therefore performed. Responses of the 19 who identified themselves as primarily working in software were isolated and then compared with those of the 34 who worked in less specialized systems.

Contrary to our initial hypothesis, the distribution of responses to the individual survey questions was very close between the two groups — so much so that even without a formal statistical analysis close inspection of the two sets of results would suggest a judgment of no significant difference.

On the surface, working with software does not provide a very different environment from that which those who worked with both hardware and software experienced. It is possible that because so many systems today are automated (meaning computerized), those working with such systems are also somewhat involved with software.

Whatever the true picture, all data — no matter how objective or how precise — require inference on the part of the researcher. In describing and particularly explaining the survey data, we make inferences and suggest explanations. In a legalistic sense, this may not be the best evidence, but it is the only evidence we have.

Specific Responses

Although the total sample consisted of 53, in some cases the total number responding to any single item in the survey was less than 53 because respondents exercised their democratic right not to answer a specific question. However, for any such items, the number opting out of that question was only one or two.

Demographic Information.   There is a broad spectrum of years of experience, with a substantial percentage of respondents having 20 and more years of experience. Most respondents had between 10 and 20 years of experience.

As is characteristic of HF personnel as a whole, the sample was almost equally distributed between those possessing a PhD (21 or 40%) and those with an MA/MS degree (25 or 48%). Only a few respondents (6 or 11%) reported having only a basic BS/BA degree.

The kind of work respondents engage in (initial design, update, redesign) is heavily loaded on initial design and update (78%). This is entirely reasonable because initial design is usually based on a predecessor system so that one can think of initial design as a form of update.

Perceptions of System Development.   The respondents perceived system development as being a formal, rational process. Intuition (a shorthand phrase for an extremely complex phenomenon) is also a major factor in system development.

Our sample sees development as a top–bottom process, but not invariably. It is also usually iterative, which suggests that later, more detailed analysis and design can force changes in earlier, higher level design (one can think of this as bottom–top). Top–down and bottom–up are not mutually exclusive and may occur at different stages of design.

Many things affect development. Most significant, we believe, is the type of design problem presented. Other nontechnical factors include: company management, resources available, and external political forces. Even in the same company, two design problems may differ in the development situation each presents.

Human Factors Adequacy.   Most writers about HFs in design have implied that the HF effort in design was inadequate as compared with the obvious need. To support this, there was the report by the General Accounting Office (1981) that sounded the same theme. Therefore, it is somewhat disconcerting for those who remember the early days to see that a substantial minority (37%) felt that HF is always or usually considered adequately in design. A minority almost as large (29%) felt that HF was rarely or never considered adequately.

One reason for this discrepancy is that, as one respondent explained, when he or she was the specialist involved with the design, HF was obviously considered adequately. Respondents who felt that HF consideration was inadequate may, therefore, have been thinking of design situations other than those in which they were personally involved. These other situations were those they had heard about, or perhaps the feeling that HF was misunderstood and inadequately treated was part of the mystique of the discipline.

Another possible explanation for the discrepancy may be that, in earlier days when HF was striving earnestly for the recognition it has now achieved (to a certain extent), it was politically correct for HF design personnel to emphasize the inadequacy of HF treatment. Obviously we still have a way to go in terms of complete acceptance, but the situation seems not to be as bad as was described in earlier writings. At the same time, the nature of the problem (in particular, whether the system is heavily dependent on the human) probably has some role to play in this question. If the system is one in which the human has only a marginal or minimal role, it is unlikely that HF will receive what some of its specialists consider sufficient consideration.

There are more and less enlightened (from a HF standpoint) engineering managements, and obviously the more enlightened will be more receptive to HF arguments.

Customer Requirements. The formal design requirements presented by the customer in a written specification or in personal interaction with designers contain information that the specialist may find more or less useful. Table 4.1 lists the major categories of information in terms of the specialist's impression of their importance. The number of respondents for each information category varied from 41 to 48.

We did not ask respondents to scale this attribute quantitatively, but rather to rank each category relative to the other categories as a function of four developmental phases: initial design, detailed design, evaluation (which is linked with detailed design), and formal testing (which ordinarily occurs toward the end of the developmental cycle). The highest rank was 1, the lowest 5. Ranking occurred within each design stage; what is presented are the median ranks for the number of respondents who considered this category.

The mission statement is obviously more important in initial design because it directs the overall outline of the design solution. It is much less important in later design, but becomes somewhat more important in testing (because testing centers on whether the completed system satisfies the mission requirement).

Equipment performance requirements are much less important than the mission statement in initial design, but become much more important in detailed design because equipment requirements are directed to the details of functionality. Moreover, equipment details directly affect the human–machine interface (HMI).

TABLE 4.1
Customer Requirements

Median Ranking of Importance of Customer Information as a Function of Development Stage

	Design Stage			
	---	---	---	---
Information	*Initial Design* *Rank*	*Detailed Design* *Rank*	*Evaluation* *Rank*	*Testing* *Rank*
Mission statement	1.0	3.5	3.0	2.5
Equipment performance requirements	3.5	2.5	2.0	2.5
Constraints	3.0	3.0	3.0	4.0
Environmental factors	5.0	4.0	4.0	4.0
Prior system information	3.0	4.0	4.5	5.0
Operator performance requirements	2.5	1.0	1.0	1.0

Operator Requirements in System Requirements Documents

Category	*%*
Always	8
Usually	40
Sometimes	31
Rarely	17
Never	4
N = 52	

Oddly, operator performance requirements are extremely important in detailed design, evaluation, and testing, but are much less important in initial design. Presumably in initial design, the focus is on the overall system, not the operator–equipment interface, which only becomes truly important in detailed and later stages.

Constraints have only middling importance. Environmental effects were consistently considered less important than other factors at each design stage. This may be because environmental effects become important only for systems designed to function under extreme conditions.

Prior system information is most useful initially and progressively loses its value as development proceeds. This is entirely logical. Once initial design has been frozen, prior system considerations are largely irrelevant.

There was also a division of opinion as to whether customer requirements contain operator performance requirements. Ideally, these would be in terms of a maximum probability of error permitted or something similar, and one hardly expects anything that precise and quantitative. However, although such requirements were uncommon earlier in HF history, apparently many specialists find some form of operator requirements presented. This may be a function of the kind of design problem addressed (perhaps the emphasis on software, which may require more consideration of users), but it is an encouraging symptom in any event because it emphasizes the human in the system.

Design Analytic Methods. What we endeavored to do in Table 4.2 is discover which of the traditional methods (so-called because they are often cited and described in texts) are important to specialists and how frequently they are employed.

The methods are task analysis (TA), operational sequence diagrams (OSD), timeline charts, decision/action diagrams, work load estimation, link analysis, and flow diagrams.

The most important of these is TA, although we do not know whether the respondents distinguished between TA and task description (probably not). TA is considered most important in initial design, but rapidly loses importance in subsequent design stages. TA is important, but is not always used.

TABLE 4.2
Importance and Frequency of Use of Task Analytic Methods

| | Importance | | |
| | Extremely | Somewhat | None |
Method	%	%	%
Task analysis	63	34	0
Operational sequence diagrams	17	65	17
Timeline charts	29	63	2
Decision/action diagrams	27	50	21
Workload estimation	25	56	15
Link analysis	10	63	25
Flow diagrams	25	60	11

| | Frequency of Use | | | | |
| | Always | Usually | Somewhat | Rarely | Never |
	%	%	%	%	%
Task analysis	29	38	19	11	0
Operational sequence diagrams	10	15	38	27	8
Timeline charts	17	33	21	21	3
Decision/action diagrams	6	15	33	27	13
Workload estimation	11	27	17	23	19
Link analysis	4	6	36	35	21
Flow diagrams	10	27	27	11	13

All the derivatives of TA listed in Table 4.2 are considered to have some value, but are used much less than TA. TA is discussed further in a later section.

Quantitative methods of comparing design configurations (Meister, 1985a) are used occasionally (24% *usually*, 43% *sometimes*), but a substantial minority (33%) rarely or never make use of the methodology. Why is this?

The procedure is tedious and requires applying complex criteria such as performance, reliability, cost, and so on. It is merely the formalization of a much more informal method of making a decision in the course of conversation. What this suggests is that highly rational decision-making methods of selecting designs are often rejected in favor of more intuitive or less formal methods.

The term *brainstorming* suggests a less than rigorous approach to design. Brainstorming is designed to elicit as many ideas as possible, most of which are ultimately discarded. It can be hypothesized as being a preanalytic methodology and is ultimately succeeded by a more formal method of analysis when it is possible to restrict designer attention to a more limited set of possible decision options. Because brainstorming as a preanalytic phase may be conflated with the later more analytic phase, respondents may not have clearly differentiated the two, and this is why there is some division of opinion about the frequency of brainstorming activity (35% usually, 35% sometimes, 29% rarely).

Design Strategies. The question addressed in this item was whether specialists and designers approached design analysis (the solution of the design problem) with some sort of general strategy (e.g., minimizing operator actions, simplification, combining individual operator tasks, etc.). Related to this was the question of whether design analysis is begun with an attempt to determine the human (and machine) functions that are required by the mission. If the if–then rationalistic design strategy were actually employed by designers and specialists, then it would logically begin with the specification of functions. Functions are initially neutral as applied to function allocation, except where they are determined by off-the-shelf components or equipment transferred from a predecessor system.

It appears that a general strategy is often used to initiate design analysis (*always*, 18%; *usually*, 43%; *sometimes*, 24%; *rarely* and *never*, 8% each). An essential part of the strategy is the specification of functions (*always*, 28%; *usually*, 51%; *sometimes*, 17%). There is always a minority opinion, and a linear, rationalistic design analysis process is not always used, although it would seem that in most cases there is an attempt to apply it. Why this process is not applied invariably is a question that needs further investigation.

Function Analysis/Allocation. HF theorists have made a big thing of FA by implication (by featuring it prominently in their writings); they have said much less about function analysis. Function analysis is simply consideration of the implications of the function for operator behavior. It may be that function analysis, which is addressed in Table 4.3, is so inherently a part of design analysis that no one thinks about it unless there is a special FA problem.

Much of the thinking about FA has suggested that it is a special process and requires special consideration apart from the design analysis that serves as its context.

If this is the case, then one must ask to what extent the specialist becomes involved in FA decisions? As in most cases, the answer is: sometimes yes, sometimes no. If there is a specific FA problem involving the human, the likelihood that the specialist will be involved in any FA decision is increased. If the design solution subsumes the human functions as part of the general design solution, probably not.

In other words, we get the impression that the specialist is not required to perform a special FA unless there is some special problem involving the human. If, in the design solution, the operator is given responsibility for certain functions, and if there appears to be no special problem with performing those functions, the specialist will not be asked by the design team to perform a special analysis of those functions.

It is possible that the design team, in considering those SEP functions that have not been preallocated, will assign them to the equipment unless there is no way the equipment can perform the functions. This can be called FA by default. Should the HF specialist urge that, if a function can be performed either by the machine or human, the human should be given preference? Some specialists may feel they should do so because this increases the importance of the human (hence of HF) in the system.

TABLE 4.3
Function Analysis/Allocation

Specialist Involvement in Function Allocation		Consideration of Human Functions in Analyzing System Specifications	Formal Function Allocation Process
Category	*%*	*%*	*%*
Always	14	82	0
Usually	39	14	22
Sometimes	29	0	26
Rarely	14	2	32
Never	4	2	18
	N = 51	*N* = 51	*N* = 49

This is where something like the Fitts list could be useful, but only if it were possible to supply quantitative estimates of effectiveness to each category in which the human is more effective than the machine. Even if both the human and machine can implement a function, the probability is that they cannot do so equally. If the specialist could show that, based on whatever criteria were applied (e.g., time required to perform the function or less design complexity), the human was quantitatively superior in effectiveness, it might be easier to convince designers to use the human option.

The specialist is also biased; in analyzing the system specification (what the customer wants from the system), the specialist almost always assumes that, unless it is manifestly impossible, all functions can be performed by the human as efficiently as by the machine.

Task Analysis.   The importance of TA in HF makes it necessary to look at it in detail. In Chapter 3, we pointed out that task description (TD) must be distinguished from TA, although the term TA is presumed to include TD.

TA is as important as theorists suggest if only because 92% of respondents use it always, usually, or sometimes, but it is not used invariably. Because it is conflated with TD, it is possible that when TA is reported as being employed *always* it is because respondents are thinking about description and calling it analysis. The task is always described because it follows routinely on the determination of functions. Analysis draws inferences from the nature of the TD and its sequencing. Analysis is important; it is used in evaluating designs (76% *always* and *usually*).

We hypothesized in developing the questionnaire that, although task description was essential, the graphic forms of TA, like OSDs, would be much less utilized because of the effort and time involved.

A significant minority (43%) agreed with our hypothesis, but a substantial majority (58%) disagreed. We have the feeling that the TA derivatives are not applied that often, but are part of the inventory of methods available to specialists if they wish; hence they are important.

User Participation.   User-centered design is all the rage among theorists these days, so we decided to find out whether the rage occurred in actual design. Theorists like Norman and Draper (1986) should be pleased to be reassured that user participation is very much involved in modern design. In fact, the percentages checking *always* and *usually* were almost 100% and by far the highest percentage found in all the questions.

At the same time, we do not know what *user participation in design* really means. For one thing, who are the users? Newman (1999) described five different users, including *operator, beneficiary, client, maintainer,* and *trainer.* We tend to think of users as the people who will operate or purchase an

SEP, but this kind of user participation may not be as common as we think. If a device is being developed for another company, are the company representatives those whom our respondents identified as users? Does user participation mean getting impressions of user desires (the attributes the SEP should have)? Does it involve user critique of initial or interim design? User-centered design is what is popularly known as a *buzzword*. What does it really mean?

What Specialists Do. We are, of course, vitally interested in what specialists do on the job. We might have given our respondents a list to check off, but the list would have been impossibly long. What such a list would have shown is that, as Meister (1999) found, specialists do just about everything, although only a much lesser number of tasks are done by every specialist.

We did ask whether specialists developed operating procedures or designed for maintenance. In times past, neither of these activities would have much occupied the specialist, but perhaps things have changed. Many specialists do develop operating procedures (79% *always, usually,* and *sometimes*), but a minority does not (17% *rarely* or *never*).

Again, in earlier days, specialists did almost nothing with regard to maintenance because it was associated with internal components. Now it appears that many specialists do design for maintenance (76% *always, usually, sometimes,* with a noticeable 25% minority of *rarely* and *never*). An explanation may lie in the fact that computerized systems like process control equipment do have as a primary operator function the monitoring and diagnosis of status displays related to system maintenance. This makes the design of displays to represent system status (important in ensuring correct performance) and symptomatology, when a malfunction is impending, a significant part of the human–computer interface and hence a specialist responsibility.

It is somewhat difficult to interpret the responses to the question about whether specialists are asked by designers to compare alternative configurations. It is logical to assume that explicitly or implicitly designers would expect (ask) specialists to compare their designs to determine which was best from the operator standpoint. Again, we have a division of opinion on this point If the point of the question is that designers request specialist services for this kind of analysis, apparently a minority (33%) have rarely or never been asked to do this. Presumably all specialists (asked or not) make such a comparison as a routine part of their work. However, if one takes the responses as literally correct, many specialists are not specifically asked by designers to do this.

In earlier days, we had assumed that specialists rarely or never performed classical (laboratory) experimental research involving variables during system development. There was neither time nor money for this, and engineering did not think it necessary. This no longer appears to be the case (28% do such research *always* or *usually*, 26% *sometimes*, 46% *rarely* or *never*). However, the definition of research, classical, experimental, or whatever, may not be too clear because we now have usability testing, rapid prototyping tests with users, and so on. Perhaps specialists are redefining what is meant by *classical experimental research.*

Rapid Prototyping. Everyone swears by rapid prototyping (RP). It is, after all, associated with the frenzy for user participation in design. The question is: Does RP occur as frequently as has been suggested? If one believes what is written about RP, there is nothing other than RP. Is it actually more effective than other design practices? These are two issues: frequency of occurrence, which is objective and can be counted; and design adequacy, which is less easy to determine.

Our respondents accept that RP is very much a part of design practice (21% use it *always*, 44% *usually*, 28% *sometimes*). They also suggest that it is more effective than other design methods (18% said RP was *always* better than traditional design practices, 47% said *usually* better, 27% *sometimes* better, and 7% said *rarely* better). Actually RP is not a design method. Instead, it is a testing method in which test results determine to some extent a needed design change. The evidence (at least if the respondents' opinions count as evidence) is not unequivocal on this point. The suggestion may be that RP is valuable, but one must keep a cautious eye on it. One can imagine a number of ways in which RP can give incorrect guidance. For example, users may wish for more than can reasonably be provided. If designers uncritically accepted those desires, they might well be led down the garden path.

A distinction must also be made between desires and needs. User impulses may focus on the wrong things — wrong for efficiency, that is. The people asking the questions in RP may ask the wrong questions or the right questions in the wrong way. Thus, the information received from RP can be confusing.

Testing. The ideal in test and evaluation in design is a formal, highly controlled process. The question is: Is it actually that formal in a design process that is often much less formal than one might wish?

Again, we get a distribution of responses that requires clarification. Much testing is *always* or *usually* quite formal (34% of the sample said so), but there are many instances when it is not (30%). OST, which occurs prior to handing a system over to a customer, is very likely to be formal. However, there may be a relationship between design stage and test formality, with testing in earlier stages being less formal. Special test facilities, as in a usability laboratory, may require formal measurement procedures. However, what often passes for rapid prototyping (simple tests or securing opinions from fellow workers) may not be very formal. Much depends on how one defines formal testing. A high degree of control is an attribute of testing, and design is not necessarily conducive to control.

Comparisons. Design naturally breaks into distinctions between hardware and software and between commercial and noncommercial (e.g., military) design. Are there significant differences in terms of the processes involved? Such major distinctions may be merely superficial.

The results we get from asking these two questions are not easily understood. Thirty-eight percent said hardware and software design were much alike, whereas 62% said not at all alike. The disparity is even greater in comparing military and product design; 52% said it was quite similar, whereas 48% said it was not at all alike.

There may be two ways to look at the design/development process. Development presents a broad picture that tends to minimize apparent differences. After all, all design passes through preliminary, detailed design, production, and testing stages. Make the picture broad enough and differences in detail tend to disappear.

If one focuses at a more detailed level of analysis, differences in design appear much more important. In software design, programming (which does not apply to hardware) is important, as are menus and icons. In military system development, rapid prototyping is less likely to be pursued simply because the user (the soldier or sailor) who is so important in commercial SEP is not that important in military hardware: Efficiency, not comfort or user desires, is more important. In commercial system development, with a great many users who have to be sold to secure a profit, RP procedures and user concerns are much more significant.

What is our point of view? The behavioral elements of design (e.g., analysis of the design problem, development of hypotheses about alternative design solutions) are likely to be the same regardless of type of SEP, and we consider these analytical traits to be more important than local differences. Behavioral analysis can accommodate itself to various situations, which consequently makes differences less significant.

Relationships With Designers. A number of questions in the survey dealt with specialist–designer relationships. This is because what the specialist is allowed to do depends — unduly so from our perspective — on designers who are engineers and almost always in charge of the design team.

In the early days of HF (see Meister, 1999), immediately after World War II, the introduction of a new discipline to engineering, a behavioral discipline, produced a certain amount of friction between representatives of the two disciplines. This created problems for specialists. The question is: Has the relationship improved?

If one can believe the respondents, the situation has indeed improved significantly. Relationships are *always* (19%) or *usually* (62%) excellent; the specialist's recommendations are *always* (4%) or *usually* (79%) accepted if they are reasonable. Of course, specialists must justify their conclusions and recommendations, but one would hardly expect otherwise.

One of the powers the early specialists fought for (and did not win) was the ability to reject proposed designs when these appear inadequate from a behavioral standpoint. If one can believe our respondents, that battle has been largely won (63% *always* or *usually* allowed to reject). Of course, the specialist's recommendations are limited to the human–machine interface.

In their area of responsibility, when testing involves human subjects, specialists are generally (63%) allowed to specify test conditions, and specialists do participate in major design decisions (63%), as one would expect if they are accepted as full members of the design team. The same is true of trade-off decisions (75%).

Specialists are *always* (37%), *usually* (47%), and *sometimes* (13%) part of the design team. The importance of membership in the design team is that the specialist becomes one of a band of brothers. If specialists are viewed as distinctive, as one of the others, it makes it more difficult to project their expertise.

Standards and Guidelines.   The data verify that standards and guidelines published in compilations like Smith and Mosier (1986) are commonly used in design, although less so perhaps in software than in hardware design. This last is because developers of software, such as IBM, Microsoft, and Apple, often prepare special style guides compatible with their software, and some software specialists prefer or are required to use them.

Some design personnel rarely if ever use standards and guidelines; for others, it is a matter of course to consult these as a check on the correctness of their designs.

Nevertheless, there is some dissatisfaction with these. A large majority feel that current standards are only sometimes or rarely adequate for design, and less than half of the respondents indicated that they were useful.

It is commonplace, a cliché, for design personnel to feel that the published research does not provide answers to the specific design problem they are facing. If all design is idiosyncratic, then it will be rare that any single piece of published research or advice will exactly match the parameters of the individual design problem. Nevertheless, the rather large investment we have in HF research suggests that the problem of the applicability of research to design guidance needs to be addressed.

System Simulation. One finds a situation entirely comparable (to standards and guidelines) with regard to physical simulations used to represent the system under design. Some specialists always or usually employ such simulations, but nearly half the sample used it only sporadically or never. Those who did not simulate probably did not do so because the type of system being designed did not warrant such a simulation — or else time, money, and effort made the simulation extremely difficult to implement.

A small minority did not find such simulations adequate for design. System (physical) simulations are perceived, however, more favorably — as much more effective than simulations in the form of models. Models are used always or usually by only 10% of the respondents; there is much greater use of system (physical) simulations. Use of computerized models may be reduced because the great majority of specialists consider them rarely or never adequate for design purposes. This is unfortunate because models have tremendous potential for design if only because systems are becoming progressively more computerized. Simulation model development is currently being given a great deal of attention in the military. There may be commercial spin-offs of military models with greater commercial use.

Information and Documentation. Information comes in various forms in design. Among the varieties are: prior system documentation, information gained by consultation with other design personnel, and information gained from HF journals.

Prior system documentation (e.g., test results) and informal consultation with other design personnel are generally found to be useful. Prior system documentation is *always* (9%), *usually* (41%), and *sometimes* (36%) useful. There is consultation among specialists and designers *always* (11%), *usually* (50%), and *sometimes* (35%).

This is in sharp contrast to material published in HF journals because most respondents do not consider this material to be useful. HF journal information is *always* (0%), *usually* (13%), *sometimes* (44%), *rarely* (38%), and *never* (4%) useful. The causes for this are explored more fully in Chapter 5.

Factors Affecting Design Efficiency. A recurrent complaint among writers on design is that time is always a problem for design personnel. Again we find a division among our respondents: 27% indicated that there is *usually* sufficient time for all the analyses needed, 33% said that *sometimes* there is sufficient time, but almost 50% said there is *rarely* or *never* sufficient time. Obviously some factor affects the sufficiency of time, but we do not know what this factor is (or perhaps there are several). The question needs to be investigated further. What is surprising is that so many respondents said there is *usually* enough time. This makes it quite apparent that time is not a completely independent variable, but interacts with other factors, such as the difficulty of the design problem.

A great deal has been written on organizational and sociotechnical aspects of the system (Hendrick, 1997). Definitions are necessary here. The sociotechnical approach is a specific way to look at how systems function dynamically, and organizational factors relate largely to management of the company units in which the individual system is developed. The question is whether, in the process of designing software or a work station, design personnel keep these overarching concepts in mind. The authors cynically believed that design personnel concentrated on the immediate design problem without considering the contextual factors in which the system was going to function. We were obviously wrong because more than half of the respondents said they did consider these factors in design. Only 11% said they *rarely* or *never* did. Of course the political correctness of considering these factors may have influenced respondents.

Software system development is often performed for institutions like banks or hospitals, which have unique organizational requirements. Developers would have to consider these requirements if the resultant software systems were to be effective.

Of course the terms *organization* and *sociotechnical* are quite abstract. This question arises: Did the respondents mean what the authors assumed organizational and sociotechnical meant? What kind of consideration did respondents give these factors and how? If they thought momentarily about these factors and then forgot them, that is one thing. If they thought seriously about these concepts, was this tied to the user-design concept, which is now politically correct? We have, therefore, another open question.

Conclusions

If it was not obvious from previous chapters, it becomes clear from the analysis of the survey results that (a) many factors, some of which are obscure, influence the design/development process; and (b) the state of our knowledge about that process is appallingly poor.

Because it was impossible to perform the long-run, focused, interactive, direct study that should be performed, the questionnaire survey, in consequence, raised more questions than it answered. Design is not monolithic;

the variability in respondent answers merely means that, without knowing the context in which design factors function, it is impossible to understand them fully.

EMPIRICAL STUDIES OF DESIGN ACTIVITIES

Methods of improving design, whether behavioral or technological (through software), must be based on knowledge of how design actually occurs. The questionnaire survey described previously was one such attempt, but obviously flawed. In the course of writing this chapter, we encountered a chapter in Moran and Carroll (1996) that presented details of how actual design groups work in early software design. Much of the material presented in this section comes from that chapter.

Olson et al. (1996) studied 10 design meetings from four projects in two organizations. The meetings were videotaped, transcribed, and then analyzed using a coding scheme that focused on the design team members' problem solving and the way in which they managed themselves.

The degree of homogeneity in design discussions that was revealed as a result of the coding surprised researchers. They may not have realized that the coding categories tended to systematize the discussions. However, Olson et al. felt that design as a task to be performed is inherently structured and to perform design at all there must be an underlying inner structure in what takes place.

Moreover, all the designers were highly experienced people. One of the two groups had previously received training on how to conduct design meetings.

The nature of the subject population, their experience, the specific training they had received, and so on are obvious limitations on how far one can extrapolate these results to other designers who are less experienced. We emphasize the experience factor because, despite this, the meetings were chaotic on the surface (the researchers' own words), which suggests that design analysis is an activity that required training. However, it is doubtful that most designers receive this training except under research conditions.

Forty percent of the meeting time was spent in direct discussion of design, with rapid transmissions between alternative ideas raised and their evaluation. Another 30% of the time was spent in evaluating progress through walkthroughs and summaries of the progress that had been achieved. Coordination required 20% of the time, and clarification of ideas took approximately a third. Obviously a great deal of time was spent orchestrating and sharing expertise among team members. There were then two general classes of activity: design and management of the design deliberations.

The teams consisted of three to seven members, and meetings typically lasted 1 to 2 hours. Everyone knew one another. The problems discussed were large and sophisticated, and involved such processes as system analysis and prototype ideas for advanced systems. The problem specification (akin to the design specification issued by a procuring agency) was vague. As a result, the design deliberations centered on further development and refinement of these requirements. The major issues focused on what features to offer a client and how to implement these features. Table 4.4 lists the categories of design meeting activity.

In one of the companies, designers had been explicitly taught a standard method to be used in design meetings, including coordination and documentation. In the other organization, such training was not given, but all designers were quite experienced in high-level design analysis (e.g., artificial intelligence problems). A variety of expertise was required that may have increased the amount of clarification required. Despite these differences, results from the two situations were strikingly similar.

The major focus was on the problem-solving aspects of design. In essence, design discussions are a form of argumentation: Various issues are raised either implicitly or explicitly. For each issue, various alternative design possibilities are presented, and eventually a decision among these possibilities must be made by applying various criteria that help select the preferred alternatives. Essentially, this is the outline of the rational theory of design (see Chap. 2).

There were also a number of other categories of activities that focused more on organizing the work of the group members. These included such coordination activities as discussion of goals, project management, and meeting management, as well as summaries, walkthroughs, and digressions. The meeting participants also engaged in clarification of their ideas. This category was defined very strictly as answers to explicit questions. These served not only to answer questions that were unclear, but also to help advance the design through refinement of the original ideas. Each of the 10 design meetings was coded; there was a high degree of interobserver correlation for the times in and transitions among these categories (range from .83–.99). Table 4.5 presents examples of common patterns and coding categories. The number of issues varied greatly from 1 to 44 over the 10 meetings. Eighty percent of the issues considered two or more alternatives, with an average of 2.5 per issue; 90% of the time was spent talking to each other.

Another empirical study was reported by MacLean, Young, Bellotti, and Moran (1996). They observed pairs of professional software designers who had worked together previously. The problem presented was one in-

TABLE 4.4
Categories of Design Meeting Activity (Taken From Olson et al., 1996)

Issue	The major questions, problems, or aspects of the designed object itself that need to be addressed. They typically focus on the major topics of "shall we offer this capability to the user" and "how can we implement that."
Alternative	Solutions or proposals about aspects of the designed object. These are typically either features to offer the user or ways to implement the features decided on so far.
Criterion	The reasons, arguments, or opinions which evaluate an alternative solution or proposal. Occasionally these appear in the form of analogous systems, with the implication that if it worked in this other system, it will be good for this system.
Project Management	Statements having to do with activity not directly related to the content of the design, in which people are assigned to perform certain activities, decide when to meet again, report on the activity (free of design content) from previous times, etc.
Meeting Management	Statements having to do with orchestrating the meeting time's activity, indicating that the group members are to brainstorm, decide (and vote), hold off on a discussion, etc.
Summary	Reviews of the state of the design or implementation to date, restating issues, alternatives, and criteria. It is a summary if it is a simple list-like restatement. If it is *ordered* by steps, it is a walkthrough, defined below.
Clarification	Questions and answers where someone either asked or seemed to misunderstand. This includes repetitions for clarification, associations and explanations.
Digression	Members joking, discussion side topics (e.g., how to get the computer to make dotted lines), or interruptions having to do with things outside the content of the meeting (e.g., discussion of why the plant was moved over by the window, or that it is beginning to snow and they should leave early today).
Goal	Statement of the purpose of the group's meeting and some of the constraints they are to work under, such as time to finish or motivating statements about how important this is.
Walkthrough	A gathering of the design so far or the sequence of steps the user will engage in when using the design so far, used to either review or clarify a situation.
Other	Time not categorizable in any of the above categories. For example, in one meeting, the group members spent time hunting within the computer for the answer to a question of whether a language could support a particular function.

TABLE 4.5
Examples of Coding Categories and Common Patterns in Complete
Episodes Involving Issues, Design Alternatives, and Criteria
(From Olson et al., 1996)

Episode 1

John:	Are we accounting for printing multiple levels of the control flow diagram?	issue
Tim:	Well, I think we need to decide that.	
	Certainly we want to print the top level.	alternative
	The control flow diagram ... George, Greg, and I thought through this a little bit, and there are some easy things to implement from a user interface standpoint for printing out selected parts of it.	criterion
	Uh, specifying the starting block and the ending block on the printing is probably the easiest way to print out sequential block numbers, if you only want to see first half of it, you can get a pretty good shot at printing knowing the first half of it.	alternative criterion

Episode 2

Don:	If knowledge editing was fundamentally collaborative then how would it be different than it is in GKR lab, that's the question.	issue
Bill:	Well, there's really two different answers to that.	alternative
	One is you want to be able to do group work which is sort of the Colab thing where you actually have people at the same time working in the same space and getting a lot of negotiation going but that isn't the one that I've concentrated most on but rather the one of how you can mediate people working together through time where basically somebody ... I mean if you look at advice, the packaging of advice in computations is really sort of a method for being able to help someone at a different time in the future. You sort of set things up in a little mouse trap sort of thing and when they, at some future date, stick their finger in the mouse trap instead of getting snapped, the cheese is	alternative
Don:	over here or something, they get help.	
	We want something that says, I knew you'd look here for the cheese but the reason you looked here is this and this and you were wrong so look over there.	
Bill:	So that's the one that I'm probably more interested in although I think the other one is very valuable.	criterion criterion

volving an automated teller machine (ATM); the designers were to critique an updated version of an ATM (a fast ATM) and to suggest alternative designs. Sessions (45 minutes in length) were videotaped as the designers sat by themselves in a room with an electronic whiteboard. Subjects were debriefed after each session.

Discussions ranged over a number of topics and tended to go back and forth in an unstructured manner. Issues concerned the order of steps in using the ATM, the buildup of queues, and why these queues occurred; and a proposal for a switchable ATM that could be operated in different modes.

These deliberations were coded using the QOC (Questions, Options, Criteria) framework that is fundamental to Design Rationale (Moran & Carroll, 1996).

Thirty-eight percent of the statements made were classified as options; it was easy to recognize these. For example, a characteristic statement might be, "Well, my favorite (option) would be just to have a 'fast-cash' button at the point where you said 'Select cash withdraw.' "

Options were usually discussed in isolation; little structure linked them. Questions in the QOC sense were rarely stated as such: for example, "Is there any way we can improve on preset amounts?" or "Do you want the receipt?"

Eight percent of the statements were classified as *issues*, which are perhaps related to questions like, "What if you asked for 50 dollars and you only had 30 dollars in your account?"

Fifty-six percent of the statements were classified as *justifications*, a very broad term that included criteria, assessments, and so on: "Well, they've speeded it up by reducing other services."

Sometimes there were trade-offs between various criteria, such as: "It is basically because otherwise you trade speed against security."

A range of general arguments, such as rationalizations, were used extensively: "If you're going to spend X amount, you might as well spend 100% and have two machines instead of one."

Sometimes analogies were employed: "You could do as they do in supermarkets, have an express line for five items only."

The reasoning structure these subjects employed can be represented as a series of questions (Q) and alternatives (A), together with criteria (C) that apply to each A. The following is only a sample of all statements made by the design team:

Q1: What range of services is to be offered?
    A1: Full range, or   C1: Service variety
    A2: Cash only   C2: Speed
Q2: How to select cash amount?
    A1: Type in amount   C1: Variety

A2: Typing and presets	C2: Variety
A3: Select from preset amounts	C3: Obvious what machine
	wants
A4: Record on card	C4: Speed and variety
Q3: Does machine hold card?	
A1: Hold card for PIN	C1: Security
A2: Return card to user	C2: Speed
immediately	

There was a consistent tendency of the designer subjects to look for evidence to confirm initial biases; this is well known in the psychology of everyday thinking. This can have the effect of the designer missing some positive design options because no positive criteria are utilized. Subjects might perseverate on poor options because negative criteria are not considered. There was no evidence of designers changing their mind about any of the possibilities they considered during the session. They never asked whether their proposed solutions had any negative consequences.

A new design emerges only gradually during a session. Analysis and innovation interacted continuously and were interspersed with other aspects of the design.

Much of the discussion in the sessions dealt with statements that justified or evaluated possible design options. Many justifications used criteria as the basis for evaluating options. Criteria are not simple well-defined entities, as some of the examples might suggest. Some criteria, such as speed, are too general for a specific design option, so one needs a way to bring the speed criterion down to an appropriate level of specificity. MacLean et al. talked about something called the *bridging criterion,* an example of which is the ease of hitting the scroll bar with a mouse. The criterion bridges between specific aspects of the design (e.g., the use of a mouse) and broad, general criteria, such as speed and accuracy. Such bridging criteria are invented for their relevance to a certain class of designs. Such a more specific criterion assumes that decisions have already been made—for example, to use a mouse and a certain kind of scroll bar. The advantage of a more specific criterion is that it is easy to work with and incorporates a more complex set of arguments and interdependencies into a single entity.

It is perfectly apparent from this analysis, which MacLean et al. expanded into more subtle details, that the superficially chaotic conceptual processes suggested by the unanalyzed design statements actually represent much more complex thinking when the surface statements are analyzed in terms of a higher level conceptual framework.

It is likely that designers are not consciously aware that they employ such complex concepts, although they can be trained to be thus aware.

Moreover, these concepts are the product of a sophisticated analysis by academics. However, these concepts can be internalized by prolonged experience with design so that designers can utilize them at a semiconscious level. Undoubtedly there are individual differences in the ability to internalize these concepts, which is why some designers are more innovative than others.

The one limitation of these empirical studies is that the design problems and the way they are presented may be somewhat artificial. Therefore, the results secured from empirical studies such as these require validation by studies in the actual working environment (i.e., the engineering facility).

The empirical design protocols already described can be further analyzed in terms of questions that designers usually ask about their design problems. These questions were derived from an analysis of design protocol studies surveyed by Gruber and Russell (1996). Our concern is what behavioral information is needed to assess the adequacy of the answers from a behavioral standpoint of does it help or hurt the operator. The following questions have been abbreviated somewhat by the authors.

1. *Requirements*

What requirements have been specified? Is a particular constraint a requirement? Can these requirements be modified in any way (i.e., how much flexibility does the designer have?)?

Behavioral implications: What is the effect of these requirements on operator functions and tasks?

2. *Structure/form of components*

What are the components involved? How do they interface? Where are they located? What are their strengths/limitations? What factors determine the choice of components?

Behavioral implications: None unless the components are involved with the user/operator in some sort of display mode required for presentation of system status information. Good design practice suggests that the user/operator should be insulated from the workings of internal components; the specialist's questions relate to whether this insulation occurs. Componentry becomes more important, of course, if we consider the human–machine interface as a component.

3. *Parameter/component interaction/failure*

How does the parameter/component behave? How does it interact with other parameters/components? If it fails, how will it fail?

Behavioral implications: Because in the new computerized systems the operator's role has become more that of a diagnostician oriented toward possible and actual failures, the specialist's concern is manifested in asking: Assuming that a parameter/component goes out of tolerance or fails, how will that failure be displayed? How visible will it be to the operator and how interpretable will it be?

4. *Functions*

What is the function of the subsystem, part, or some feature of the part?

Behavioral implications: The human in the system can be thought of as a subsystem, and so its function must also be considered, along with those of the physical components. Inevitably this involves a consideration of how the physical function affects the behavioral one.

5. *Hypotheses*

What happens if the parameter changes to a new value?

Behavioral implications: The new value may or may not (this remains to be determined) have an effect on what the user/operator is supposed to do in the system. This is especially so if the operator functions as a diagnostician of what has gone wrong.

6. *Dependencies*

What are the dependencies among the components and subsystems? Is any parameter particularly critical (this may act as a constraint on the system), thus dominating others?

Behavioral implications: Again, the dependency may cause a failure or out-of-tolerance situation to which the operator must pay attention.

7. *Constraint checking*

Are known constraints satisfied by the design? Does the system, as now designed, satisfy known constraints?

Behavioral implications: Constraints include human thresholds. Engineers are not likely to propose physical constraints that unduly stress operators, but they may not take sufficient account of cognitive constraints (flow of displayed information, information quality and quantity, etc.).

8. *Decisions*

What decisions were made with regard to functions, subsystems, components, parameters? Where did the idea for the decision arise?

Behavioral implications: Decisions are cognitive factors. The specialist's concern is whether the decisions made have any significant impact on what the human in the system is expected to do.

9. *Justifications and evaluations of functions and alternative designs*

Designers experience a great need to feel that their design decisions are rational and valid. Therefore, they will express their concerns during design team meetings. Functions are required by the logic of design (e.g., detection involves monitoring). Alternative designs are also constrained by logic, but less so because there are different ways in which a function can be implemented (e.g., monitoring can be accomplished by visual or auditory mechanisms).

Behavioral implications: Specialists will be concerned that both the positive and negative consequences of these decisions be explored, with particular reference to the system operator.

10. *Validation*

Designers, recognizing their great responsibility, seek ways to assure themselves that their decisions are justified. They ask: How is a particular requirement satisfied?

Behavioral implications: The specialist is no different from the engineer designer in seeking justification of his or her activity during design. Because the specialist lacks full responsibility for the completed design, he or she may lack the same degree of anxiety the designer feels. If logic were the only factor in design, the designer might not feel overly anxious. However, because intuitive factors enter into design, they are a source of anxiety and must be exorcised by whatever logical rationale can be produced. Nonetheless, as has been pointed out previously, decisions made intuitively may be based on logical processes performed subconsciously.

Designer anxiety may also be induced by fear of overlooking something that might be obvious to someone else (i.e., an error of omission).

An Aerospace Study

Special attention should be paid to a study by McCracken (1990), which explored the question asking and information seeking of aerospace crew system designers. (The relationship to the QOC approach of Design Rationale investigators should be noted.) The purpose of the study was to provide the basis for developing a decision support system (DSS) for designers, but our concern is not for the DSS, but for the conceptual processes explored. The following is taken from McCracken's Ph.D. dissertation, which can serve almost as a model of how research on designer analytic processes should be conducted.

This study, which was performed over 21 months, involved seven aerospace crew designers. (Questions can be raised about the representativeness of such a small, specialized sample, but this limitation is characteristic of all research subjects other than college students.) The seven subjects worked at/for the USAF at Wright–Patterson AFB, as did the investigator. Interviews and integration meetings were conducted, audiotaped, and transcribed; designers kept diaries, and the researcher participated as a member of the design team. Documents produced by the team members were collected and analyzed.

The engineer's knowledge structure took the form of a hierarchy for the function analysis, which was a large part of the design team's work. A major finding was the existence of high-level questions developed by the team leader that guided the design synthesis process and organized the question-asking behavior of the other design team members.

One of the questions examined in the study dealt with information-seeking behavior. Allen (1977) described that behavior in engineers as being in inverse ratio to the effort needed to collect the information (see also Zipf, 1965). Hence, engineers relied principally on their own knowledge and that of fellow workers (see the corresponding response of the survey respondents).

McCracken found that information was needed in relatively short response cycles, and the information had to be contextualized for the type of crew subsystem being designed. (This is the common difficulty of specifying and applying more or less general information to a specific problem's questions.) This means that overly general information cannot be applied very well (which may explain the lack of utility of HF research literature and standards). Almost no use was made of computer sources, which suggests that if the DSS is computerized (as it probably must be), it must take a form different from what is presently available.

There were a number of problem-solving strategies that must be examined more closely. These included:

1. *Island hopping*. This phrase describes the strategy of working on one dimension of the design problem and then changing to another dimension, such as going from developing the function analysis to doing a technology assessment based on that function analysis. There were two reasons for this, as reported by engineers. The first was to alleviate a feeling of burnout with a particular form of the work. The second was external to the subject; either some resource needed to perform the work (a meeting room, a computer) was not available or some authority figure scheduled an activity that took priority.

2. *Boundary or constraint finding*. This was the designers' determination of certain limits within which their design would evolve. These limits included time, schedule and budget, hardware availability, engineering support, software development, and manpower availability. These external forces changed during the course of design and had to be sampled iteratively to maintain awareness of boundary conditions. (It should be noted that, in a complex design situation, the problem of situation awareness may sometimes arise; so many facets may occur simultaneously that designers may lose track of some of them.)

3. *Crossword fill-in*. This phrase, developed by McCracken, is somewhat difficult to describe. The complexity of crew system design is such that so many variables and interactions are involved within a domain that it must be divided into subsections to be at all manageable. Designers were forced to create a structure within these subsections by identifying a

known structure that hinted at what the surrounding structure would look like. Development of a priority subsystem to aid in a format paging system is an example: During air-to-air combat, the pilot is not going to want landing instructions, so it is necessary to provide him with weapon system status information, but what priority should this displayed information have? By identifying some structure—probable need within the flight phase and a logical prioritization scheme—designers could then proceed. McCracken saw this as akin to a crossword puzzle, in which filling in one word provides clues to others and delimits what those words can be; only certain words can be used that match the ones already filled in.

4. *Hierarchical decomposition.* This used the function analysis that was apparently an initial organizing structure to break the knowledge domain into more manageable chunks. The hierarchy was as much as seven levels deep depending on which system was being addressed.

5. *Peaks and depths strategy.* A top-level decomposition scheme across the major functional areas was completed first. This involved the following structure underlying most of the major function categories: (a) determine present state of the system; (b) determine desired state; (c) determine the difference between (a) and (b); and (d) decide the action to be taken. The design team exhaustively decomposed a major functional area before proceeding to another (of course, areas were revisited as more was learned; however, this was not a conscious strategy as such).

6. *Analog structure.* This used previous analyses that seemed to apply to new problems (e.g., a function analysis performed for a prior bomber design problem). The exact structure of the prior function analysis was reworded, but some pieces were imported almost intact.

Because the crew system problem was quite ill structured, more questions were focused on generating structure than on seeking specific design information.

In a team situation, the leader gave assignments to various members to perform parts of the analysis (e.g., design scenarios, function analysis, technology assessment; what is the status of technology with regard to some aspect of the crew system under consideration?). He requested periodic updates of what was being done in intervals between team meetings. Members had been divided into design (analytic) and development (production) groups, which necessitated each briefing the other.

The design process observed by McCracken was evolutionary, involving iterative cycles. His cockpit design was second generation, which highlights the fact that even in second-generation design the problem may be ill structured and require definition by questioning. McCracken argued for the recording of design histories because many questions are generic.

This also accords with the philosophy of Design Rationale (Moran & Carroll, 1996).

Given knowledge of the required functions, information, and technology available, as generated by system analysis, and knowledge of the user's job (the function analysis), the designer can begin to aggregate the components.

McCracken saw this as the creative synthesis part of the design process, performed largely in the head of the experienced designer and guided by high-level questions he asks. These questions acted as a short-term memory aid by chunking problems at a level that combines variables and relations.

It appears that at least one knowledge-based system for design (Coyne, Roseman, Radford, Balachandran, & Gero, 1990) was in process when this study was performed.

Research into designer practices and, in particular, their concept structure has so far consisted of highly structured problems in what amounts to classroom settings. This may be adequate as far as it goes because it is assumed that in real working life engineers attempt to structure the way in which they solve design problems. However, this kind of research describes the design process as a series of immediate, direct, face-to-face design team meetings. The design tools available to the specialist and designer may also require some other period of time during which they perform detailed analyses of the problem. Whether that time is available is unknown because half of our previous survey respondents reported that they lack the necessary time to perform their functions adequately.

Moreover, design takes place in an enveloping contextual process called *system development*. We lack the knowledge to determine how much of whatever goes on in that development impinges on and distorts the picture presented by research of highly structured settings. What goes on between design team meetings? To what extent are the specialist's design tools like TA actually utilized?

The commonly described design tools are the products of abstraction. Just as the Design Rationale is the ideal way in which designers should analyze the design problem, so methods like TA or the TA derivatives represent the ideal way in which one should analyze the design problem from the specialist's standpoint. Abstraction, by its very nature, creates a somewhat artificial product. R. Miller (1953), who wrote the first manual on TA, almost certainly did not write it after conducting research in an engineering facility and observing how the HF specialists of that day went about their work and their analyses. Even if he had made such observations, the position of HF personnel in engineering in the early days was sufficiently anomalous (Meister, 1999) so that there may be substantial differences in how they pursued their work then and how we do it now.

Moreover, the later analytic techniques available to the specialist were probably also not developed on the basis of empirical research.

It is quite frustrating to realize that the task-description process we apply in user-centered design has not been applied to the study of specialists and designers in their working environment.

5

Information Resources

We have postulated that design is an information processing activity centered around questions that must be answered to achieve design goals. Those questions were described earlier.

The kinds of questions for which answers are sought are different for designers and HF specialists, although they overlap, as they must, if design is to be adequate. The information that designers seek is related to questions of functionality (e.g., to produce a particular effect, what kind of component, functioning in what way, is needed?). The kinds of questions specialists ask are related to the effect of the way in which functionality is achieved (i.e., the effect of that component and its functioning on what the operator must do to perform adequately). In the case of the specialist, there are actually two major questions: Is a specific system factor/characteristic important in terms of its possible effect on the operator? What is the nature of that effect? (Of course, if the specialist is given responsibility for design of the interface, he or she will also be concerned about the functionality of that interface.)

Logically, the designer should also be concerned about the effect of equipment functioning on the operator because the operator is a major subsystem of the system being designed. However, because the engineer is concerned primarily with functionality and is not a behaviorist, he or she assigns the operator responsibility to the specialist. In worst-case situations, the designer is not even aware of the operator, and the specialist must harass the designer to ensure that personnel factors are considered.

Answers to the questions that are asked come from various sources that are often the same for both the designer and specialist. These sources are general and specific. The general source is research performed in the past, the results of which provide principles and data published in textbooks, manuals, and journals. A more specific source is the accumulated experience of the individual designer and specialist. Still another specific source is the experience of other designers and specialists to which any designer/ specialist can refer. Allen (1977) found that a primary information source for engineers is the expertise of other engineers. The reliance of the individual on other members of the design team is why the design meetings illustrated in Chapter 4 are so important. Design is not a solitary process (unless one is perhaps an Einstein) because no single member of the design team feels sufficient confidence in his or her own judgment to go it alone. Even Edison, arguably the greatest American inventor of his time, relied heavily on his design team.

Answers to design questions are also produced by if–then reasoning, but design is not completely or solely a matter of logic, and certainly not for the specialist. Facts may serve as propositions (e.g., if, as research tells us, the effect of system factor X on the operator is thus, then, by inference the operator's performance will be affected in the following way).

The designer's experience with previous design problems for which solutions were achieved is extremely important, at least as far as suggesting a particular strategy with which to attack the present problem. As we have indicated previously, this may be the theoretical basis for the creative and intuitive elements in design.

Experience is somewhat less important to the specialist because he or she is less concerned with solving design problems as such (i.e., functionality) than with determining the implications (effects) of any possible solution on the human. The engineer has principles such as those of electrical flow and strength of materials, which serve as a primary information source for the design solution. In the same way, the specialist has (or rather should have) accumulated HF research to serve as an information base. Unfortunately, that research is weak in terms of providing information relevant to a specific design problem.

When research is unavailable, or the data are weak or ambiguous, specialists cannot make unequivocal statements about the effect of a system characteristic on the operator. The best they can do is develop a hypothesis of what the effect will be. That hypothesis, which may be based on knowledge or experience, is affected by what we call *analogical reasoning* (AR). We define AR as an identification of the specialist with the potential operator/user of the system, as a result of which the specialist substitutes for the missing or ambiguous data his or her knowledge or feeling of how the specialist would react to the situation at issue. This replaces the missing or ambiguous knowledge with the hypothetical reaction. More concretely, AR makes use of the following logic: I am a human; if I were confronted with the following system stimuli, I would react in such and such a way; therefore I reason that the average operator (whom I resemble) would react in the same way. This reasoning is rarely that deliberate and conscious (the specialist has been trained that this kind of reasoning may be biased or fallacious), but at a less than conscious level it will have an effect on his or her conscious reasoning processes.

In searching for a design solution, engineers do not use analogical reasoning because they are concerned with physical properties and processes with which they cannot identify. Of course, when the designer thinks about the human role in the system being designed, he or she is also subject to AR reasoning.

AR exercises a particularly potent effect when the source of the information has been produced by direct contact with people, as in interviews with operators, observation of personnel performing their duties, questioning of subject matter experts (SMEs), and so on. The intimacy of the contact fosters the unconscious use of analogical reasoning. There is no such intimacy when one is reading a textbook.

Just as the designer creates hypotheses that are examined as design options, so the specialist creates hypotheses about the potential effect of a system characteristic on operator performance. In a peculiarly paradoxical way, even the specialist's failure to develop a hypothesis about a system effect on the operator represents a hypothesis; it is that an effect will not occur. Unless research knowledge is unequivocal, as with human thresholds (e.g., an operator cannot move a 500-pound weight without mechanical assistance), the hypothesized effect of a system factor must be tested, and AR may serve as the test mechanism, which will result in acceptance or rejection of the hypothesis.

AR is an inferior kind of logic because it relies on a feeling of identity between the specialist and the operator about whom the specialist is predicting. We say AR is inferior because it compensates for an unavailable objective knowledge, which is what HF research should (but sometimes does not) provide. Because human behavior extends over a range of performances, the significance of the effect of any single performance (except at the threshold level) can only be hypothesized, which is when AR takes over.

AR also exercises an effect when it is necessary to apply general research to the specifics of the design problem because this application always occurs with some uncertainty. The extent of the applicability is an unknown, which produces uncertainty. AR helps to resolve the uncertainty and does so unconsciously, which is why AR is so insidious. Because I am human and I would react thusly, so the operator would react . . . ; if all this sounds like the thinking of sympathetic magic, it is. Facts do not lie, of course, but the interpretation of these facts takes place in the context of AR.

We have gone to such lengths to explain the nature of AR because its existence highlights the need for objective HF research data that are also applicable to design. The effort to understand the information-seeking and information-processing characteristics of the specialist and designer is important because whatever the formal outputs of the design process (e.g., blueprints, scenarios, software programs), these are only the outward expression of largely mental activities. It may seem strange to say this about engineers because they deal with physical principles and artifacts, but the essence of design (and also what the specialist does) is cognition. No effort to improve design quality can be effective if it does not recognize this essential feature.

As far as the specialist function is concerned, its efficiency in design cannot be increased unless the research on which it depends is improved. This is why it is necessary to examine the utility of research literature.

Information of varying types can be used to facilitate HF design. Two general types of information are available: design-specific and contextual. One form of design-specific information is derived from what are called *HF design guidelines*. These are derived from HF research, common sense, and logic, and they specifically relate to the design aspects of the human–machine and human–computer interfaces (HMI and HCI). This information contains prescriptive aphorisms or commands such as "Scroll up, not down" or, "Make all dialogue language consistent."

Another form of design-specific information is engineering data about internal system components that must be transformed by HF design specialists into interface controls and displays.

Contextual information includes the task description and task analysis (TA) and information from predecessor systems to the one being designed. An example would be an earlier copier machine. This information, which describes how personnel of the predecessor system performed their tasks, is derived by examining operating manuals and test data reports, through interviews, questionnaires, surveys, and even physiological and physical measurements (Luczak, 1997).

Contextual information from TA and predecessor system information is valuable because it provides the context in which design must function. It provides a framework for evaluating new design adequacy and how the new design will be utilized.

The term *design guideline* is inappropriate because rather than guiding design, such guidelines are useful primarily to evaluate the adequacy of an already created design. The most common use of guidelines is to compare the newly developed design with the guideline prescriptions to ensure that there is no conflict; if there is, it is necessary to correct the new design.

A simplistic example can be used to illustrate guideline usage. The guideline says that the most important and frequently used controls and displays should be located in a central position on the workstation. If the specialist, checking the workstation design against the guideline, finds a maintenance control and display, which is used very infrequently, in the most prominent position on the control panel, it should be moved to a more peripheral position.

In this example, other reasons may be found in the TA for the maintenance control and display to be positioned as it was. The guideline will then be ignored for this design. Contextual information is necessary because, if the designer did not have the task context to illuminate the use of design, the design may be inappropriate.

Design-specific engineering information describes: (a) the internal componentry of the system being designed; and (b) information about the external environment or objects outside the primary system. In developing an HMI or HCI, information must be made available to the operator that will tell him or her whether the system is performing adequately. This provides the operator with clues (symptomatology) about the status of important internal components. For example, information about steam pressure in a nuclear power plant boiler must be provided so that deviations from preset pressure levels can be noted and emergency actions taken.

In systems like radar and sonar, devoted to recognition of external world events, operators must receive information on the interface about those events. For example, the interface must provide information such as the arrival in the vicinity of one's own ship of a possible enemy submarine.

The engineering information gathered from engineers includes the following: which components are the most important for safe functioning of the system; what performance characteristics of these (symptoms) reflect their status; what the effect will be of deviant functioning of these components; and what actions both in the short and long term the operator must take in the event of a malfunction.

The judgments of HF design specialists (with the aid of engineers) will determine which information, and in which form, shall be presented to the operator of the interface. It is with regard to this last point that HF design guidelines are useful to assist in the arrangement of interface controls and displays. Unfortunately, HF design guidelines are limited; they have little to say about avoiding information overload, the way in which separate items of information should be combined to facilitate interpretation, and so on. These guidelines focus on relatively molecular aspects of the interface; they have difficulty in providing requirements involving higher order constructs dealing with information utilization and analysis, which, because of the increase in advanced systems, are becoming more common.

The nature of the task is supposed to direct interface design, and the task does have significant effects on the design because the design exists only to implement task performance. Even so, there is room for considerable flexibility in deciding what information and how that information should be presented. For example, if the task requires the input of certain information to the system, the input can be accomplished via keyboard, light pen, menu selection, or voice-recognition command. This flexibility gives the designer and specialist a number of choices. The essence of the design process is the use of information to reduce the number of choice alternatives until only one or two are acceptable.

The reduction in the number of design choices is facilitated by the application of criteria that are accompanied (one hopes) by design-specific and contextual information. These, together with task description and TA information, reduce design choices by suggesting that one alternative design choice is preferable to another.

Design constraints are provided by criteria such as: Do not overload the operator's information-processing resources, provide only that information which the operator requires, and do not require the operator to perform overly complex and time stressful perceptual-motor and cognitive activities. These are admirable criteria, but relatively useless unless concrete quantitative data accompany each criterion. For example, in this particular design situation, which is the necessary and excessive information? What is overly complex activity?

There are two problems here: first, the definition of high-level constructs like information overload; second, the application of these constructs to the specific design problem. It is theoretically possible to provide general guidelines about information, but such general guidelines must be tailored to the specifics of the design situation. No set of guidelines can be sufficiently specific unless research is performed specifically for a particular design situation. The research by Campbell and his associates (Campbell, 1998; Campbell, Carney, & Kantowitz, 1999) for the Automatic Traffic Information System (ATIS) has demonstrated that one can develop quite specific and hence useful guidelines. However, most design projects do not have the funding to research and develop such design-specific guidelines.

In this event, the specialist in particular must take whatever information is available and tailor its use to the specifics of the individual design. The resultant initial design is then exposed to a number of evaluations and tests, including (a) comparison of HF design guidelines with the design; (b) a walkthrough, which is a behavioral simulation of the use of the design, with the TA establishing the context in which the design is used;

(c) exposure of the design to subject matter experts (SMEs), who make evaluative judgments of design adequacy; and (d) prototyping tests of the design, including, as part of the prototyping, human performance tests of the capability of operator personnel to utilize the human/system interface.

In the last item, the sample interface is presented to subjects dynamically on a computer terminal. Subjects are presented with sample problems of symptomatology displays to analyze, interpret, and diagnose in much the same way that expert systems present medical information and diagnoses to physicians. Response adequacy (e.g., 8 of 10 problems correctly diagnosed) would determine whether the proposed interface does or does not permit operators to perform their jobs adequately. Any design not satisfying the pass–fail criterion would be modified and operators retested with the changed interface.

HF research has been criticized because the guidelines it provides are not completely satisfactory (see Meister, 1999, for details of this). Because the guidelines are abstractions from the original research, they can never be fully satisfactory in their application to specific design. Part of the criticism is that too little HF research addresses questions pertinent to design. As seen later, the guidelines available deal mostly with molecular design aspects and, in particular, lack material dealing with cognitive functions.

What do we do about all this? We can improve design guideline information by expanding HF research to address applicable design issues.

Information must not only be available, it must also be easily accessible. Special efforts must be made by specialists to secure access to predecessor system data because this may not ordinarily be transmitted at the start of design.

It should be a recognized responsibility of HF design specialists to begin early in design to gather engineering information about internal components in preparation for interface design. There seems no reason to suppose that engineers will be reluctant to supply this information. HF design guidelines are ordinarily found in reports, journal articles, and books, of which the most well known are Woodson (1954), Woodson, Tillman, and Tillman (1992), and the Engineering Compendium of Boff and Lincoln (1988).

To be quite up to date, publication includes not only paper copy, but also electronic media (Budnick, 1999). This permits wider exposure of the specialist to ergonomics material using powerful computer search engines. What is useful in the new medium depends on the relevance of the original material entered into web sites, however.

Sources of potential information exist in scattered papers and in many journals. HF specialists and designers need help to access and integrate that information. That means that personnel (it is almost never a one-person job) must assemble the individual papers, extract the useful part of each paper, combine these with other related items using some sort of taxonomic organization scheme, and develop a summary statement (e.g., a handbook) that in essence says to the reader, here is everything you need to know (so far) about X.

The first successful design handbook was Woodson's (1954) *Human Engineering Guide*. Since that time, a popular conception has developed that the publication of an improved HF design guide would be something like the Winchester rifle, which subdued the west. In HF terms, such a document, if it fulfilled expectations, would answer the many factual questions about operator performance that arise during design and help create a positive attitude toward HF in engineers.

Our concept of the design process is that design is organized around questions to be answered, and much design activity is the effort to secure the information needed to answer these questions. This activity involves accessing sources other than one's own information and memory store. In the course of their experiments studying the value of information in design, Burns and Vicente (1996) developed criteria of the effectiveness with which information is accessed: *cost*, in terms of time and effort to find the answer to a question; *importance* of the answer to the design question (i.e., in terms of what is necessary for a design to be implemented); *relevance* of the information; and the *effort* design personnel are willing to expend to find the answer to a question.

Leaving aside personal experience and the experience of others, that information should come ultimately, if not immediately, from HF research. We say *ultimately* because the original information may have been learned from reading journal articles or textbooks by one individual and then passed on verbally to another, or it may be encapsulated in something other than a research report (e.g., in an engineering report).

THEORETICAL CONSIDERATIONS

Our concept of HF as a unique discipline assumes that the research produced by it should be the bedrock from which answers to HF design questions are derived. Burns and Vicente (1996) listed a number of ways in which the designer can secure information: (a) consulting operators or users; (b) conducting a formal or informal experiment; (c) already knowing the answer; (d) asking a colleague or use personal judgment; (e) consulting a journal article the designer already has; (f) looking for a relevant journal article in a library; (g) accessing a computer-based database; (h) using an analytical model; or (i) looking up a previous or similar design. For most of these methods of securing information, previous research is a requirement. If HF research has any purpose other than to provide post facto explanations of phenomena (a purpose that we consider to be insufficient for a pragmatic, development-oriented science, which HF is), that research must be usable and used by HF practitioners, if not by engineering designers.

Just as machines must have functionality (they must be able to perform assigned functions unless they have malfunctioned), so research dealing with human performance in relation to machines must have functionality, which means that the research must tell us something that will enable us to design machines more effectively for human performance. As an example, if one does a study on aging (e.g., Mead & Fisk, 1997, 1998), showing that the aged do less well with automated teller machines (ATMs), the study has little value unless it also tells us how to design an ATM that will allow the aged to use it more effectively and will possibly motivate them to use an ATM.

For many years, starting from the seminal work of Meister and Farr (1967), grave doubts have arisen concerning the value of HF research because there is a division of responsibility between the HF design specialist and the researcher. In the words of Burns and Vicente (1996),

> [h]uman factors engineers are . . . responsible for applying the knowledge of human physiology and psychology in the design of objects. . . . In contrast, the human factors researcher is responsible for expanding the knowledge and *transferring* that knowledge to the human factors designer. (p. 259; italics added)

The distinction Burns and Vicente made between HF engineers and HF researchers mirrors a gap that Meister (1985b) referred to as the "two worlds of human factors." As Burns and Vicente pointed out, the problem is not that HF information is unavailable (indeed, for many years there has been a small cottage industry of developing HF handbooks to be used by designers), but that much of this information has had little influence on design (Vicente, Burns, & Pawlak, 1993). There have been those who have attempted to solve this problem by computerizing the information (Monk, Swierenga, & Lincoln, 1992; Whitaker & Moroney, 1992), but this solution (which is part of the Zeitgeist that sees computerization as the solution to all problems) is unlikely to solve this particular problem. Computerization is a medium for presenting information; it does not create information on its own, although in all fairness it must be said that sophisticated computers can integrate disparate information items and, on the basis of an algorithm, extrapolate new information.

A number of researchers (Cody, Rouse, & Boff, 1993; Rouse, 1986; Rouse, Cody, Boff, & Frey, 1990, 1991) have suspected that more fundamental problems underly the inadequacy of HF research for design. One of these problems is the failure to study design as a behavioral process. In particular, Rouse significantly advanced our thinking about design as behavior by developing a model of the designer's information search criteria (Rouse, 1986). Rouse concentrated on the value of design information, defining *value* as that which the designer is willing to pay for in money and, more consciously, in effort.

The designer's use of information is based on the perceived value of the information and not on its objective properties (this is an example of how design is quintessentially a behavioral process). Information is accessed only if the designer perceives the value of the information to be greater than the perceived cost of securing it. Certain dimensions affect this perception of value: the reduction of the designer's uncertainty with regard to a design question, the relevance of that information to the design question, and the appropriateness of the form in which the information is packaged. Rouse defined *uncertainty reduction* in terms of gaining knowledge of a previously unknown fact; appropriateness of form is measured (negatively) by how little transformation must be applied to the information to make it useful.

One must also differentiate between information that is relevant and necessary and information that is relevant and discretionary. The latter can be ignored (nice to know, but . . .); the design problem can be solved without applying discretionary information. An example might be details of the Helmholtz theory of color vision; this may be relevant to the problem of selecting colors for computer graphics, but the selection can be performed without the theory.

Within information science as a whole, there is general agreement that information relevance is a dynamic state depending on the designer's perception of that information at a particular time. In other words, information that is relevant at Time T may not be as relevant at T + n. What changes between T and T + n is perhaps the specific design question that needs answering.

The factors we have been talking about have been formalized into a theory developed by Burns, Vicente, Christoffersen, and Pawlak (1997), which is an adaptation of work by Pejtersen (1985). That theory is represented in Figure 5.1.

Research findings are primarily summarized in handbooks, but also in individual research papers; those who write handbooks (e.g., Boff & Lincoln, 1988; Woodson, Tillman, & Tillman, 1992) act as intermediaries or transmitters; they select information on the basis of perceived relevance and package it to be used by design personnel. The problem they face is that the information was gained originally in a specific context, which may not be the designer's context in retrieving the information.

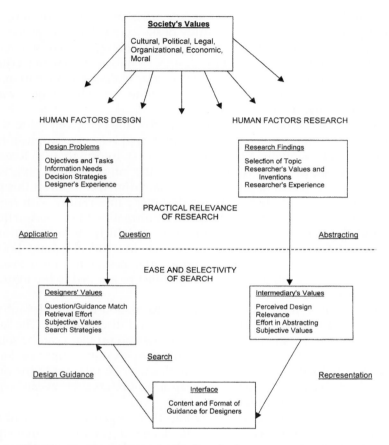

FIG. 5.1.   Framework for understanding the factors impacting the transfer of HF research findings to design (from Burns et al., 1997 with permission).

Burns and Vicente (1996) and Burns et al. (1997), from which much of this section has been adapted, have performed a number of studies in which they asked subjects to rate the relevance, importance, cost to obtain, and efforts they would make to secure the answer to a design question such as, With a constantly monitored display, is an increase or decrease in parameter value detected sooner? Thirty-five design questions sampled from Boff and Lincoln's (1988) Engineering Design Compendium (EDC) were sent to 40 HF nuclear control room designers. Twenty responded; 18 responses were usable. Results of the study indicate that importance and relevance were highly correlated. HF handbooks were considered a relatively low-effort source, with a mean of 25.8 on a scale of 0 to 100. There was a consistent tendency for designers to overrate difficulty of securing information. The two most effortful methods were considered the formal and informal experiments, with the formal experiment being rated approximately 98 on the 100-point scale. Although handbooks were relatively low-effort resources, the information they provided was rated relatively low in importance.

HF information is not considered very important by engineering designers, a conclusion that was anticipated earlier by Meister and Farr (1967), although that situation has probably improved since then. This suggests that the HF discipline must make strenuous efforts to provide information of high value to design, make the design implications of research clearer, and create information that more closely matches the design context of the information users.

Burns et al. (1997) also did a number of studies comparing the EDC with *Human Factors Design Handbook* by Woodson et al. (1992). The two handbooks represent diametrically opposed conceptual frameworks — the Compendium following an academic information-processing model, the Handbook being focused on design domains and problems. Three types of subjects were utilized: engineering undergraduates with no HF or design experience, engineering undergraduates exposed to HF in their class work, and professional HF designers.

Using either or both handbooks, subjects were asked to answer four questions, answers to which could be found in both handbooks. A sample question was: What is the quantitative relationship between ambient noise and speech intelligibility? Subjects were asked to locate the answers to the four questions (25 minutes allowed per question) and verbalize their search goals and methods. Results indicate that subjects answered 80% of the questions correctly from the Handbook, as opposed to 56% from the Compendium. Search time for the Handbook was an average of 7.1 minutes, whereas it was 10.5 minutes for the Compendium.

In another experiment, subjects were asked to rate 35 design questions taken from the Compendium on the four dimensions cited previously of cost, relevance, importance, and effort. Most of the ratings of the design questions were low, suggesting that subjects felt the questions had low relevance and importance. Subjects were also asked, as in an earlier study, to rate the venues of the information and sources noted previously. Handbooks received relatively low effort ratings. Perhaps designers feel that HF design information is generally of low value. However, in a subsequent field study, designers did identify certain questions that were important to them. These questions stemmed directly from the nature of the design problem. Unfortunately, answers to these questions were not found in either the Compendium or the Handbook because of their great specificity. Manifestly, only that HF information that is specifically tied to immediate design questions is considered relevant by designers.

It is possible to make HF information usable to designers, but it seems unlikely that such material could be extracted from the general behavioral literature, which is largely uninterested in design nor stimulated by design questions. Rasmussen, Pejtersen, and Goodstein (1994) suggested a strategy for developing a useful HF database:

1. Study the needs of the specific design domain;
2. Study designers' strategies and criteria for information retrieval;
3. Identify areas in which information resources do not match demands imposed on designers;
4. Develop a new design aid, tool, or method suited to the nature of the problem;
5. If the solution appears to be a handbook, search the literature relevant to the design domain of interest; and
6. Make the handbook usable, legible, and comfortable.

This is a rather idealistic strategy because each step requires a great deal of research, for which enthusiasm in the HF community seems lacking. However, it is realistic in suggesting that the first thing one has to learn is what the real information needs of the specialist and designer are. This makes it necessary to study both types of HF information users. The concept of matching the available HF data against what is needed is a prerequisite to the performance of additional design-focused studies. The unfortunate aspect of these recommendations is that it is unlikely that the HF research community will act on them.

The fundamental question to be answered is: Is it possible to distill from HF research literature (which is largely unrelated to design) the particulars that specific design problems demand? Design problems are highly specific; that is the main reason behavioral specialists engage in a search for relevant information, wherever it can be secured.

This is the rub: To secure HF information of interest/relevance to engineers with specific design problems, one must perform the research specifically to secure that information—that is, one must perform design-problem specific research. When this is done (e.g., Campbell, 1998), the design guidelines resulting will be both relevant and important (and accepted by engineers). However, then one has to fund such specific efforts, and it is doubtful whether more than a few high-visibility design domains will be thus supported.

Moray (1994) suggested that, because design is highly context-specific, it is impossible to develop HF generalizations in the form of guidelines.

Nevertheless, even granting the specificity of design, it should be possible to develop classes or types of system contexts based on the commonality among the functions personnel must perform (and the relative importance of each function based on a scale that requires development). For example, there is a class of system in which personnel must read and integrate data from varied sources. There is another class of system in which personnel must monitor infrequent, weak signals that occur at unpredictable intervals. One faces the problem of having to combine these functions because few systems require only a single function. However, combined functional contexts can be taxonomized, and then it would be possible to study the characteristics of these systems.

We cannot accept Moray's (1994) overly pessimistic view. That is because there are already guidelines that can be applied in a variety of design situations (examples are provided later). We believe that it is possible to develop a middle way between complete research generality (which would indeed be irrelevant to design) and HF research directed solely to special design interests (e.g., research to provide guidelines for nuclear power plant control room designers, designers of automated traffic navigation systems, etc.). Specialists other than those special users for whom the research was specifically performed would have to extrapolate the information in these guidelines to their design problems; much of the information in these special guidelines would be of little value to them.

Our feeling is that, as indicated previously, design problems fall into a limited number of descriptive (taxonomic) categories and that the information needed to provide behavioral guidelines for these problems also falls into a limited number of categories. The development of these categories will require much thinking, the output of which will be one or more taxonomies around which relevant research can be structured. We consider the problem of meaningful design-relevant research as solvable, but it will demand will power and a systematization of research that some will consider dictatorial. That is, after we have determined what the design information needs are and the nature of the research required to respond to these needs, researchers must be induced to provide the needed information through their research.

Researchers will in general study whatever they can get funding for; this is cynical, perhaps, but true. Because only government or entrepreneurs like Bill Gates (Microsoft CEO) have money these days, it will be necessary to cajole one or the other (or both) to put up the cash; then the problem will be on its way to solution.

Of course, technology changes, which means that the research effort must be a continuing one. Our philosophy, however, is to take the effort one step at a time.

RESEARCH ON DESIGN GUIDELINES

The HF design specialist has as his or her resources: (a) experience, including training; (b) relevant HF literature as he or she interprets that literature; and (c) the distillation of that literature in the form of guidelines (principles and recommendations) and standards. This chapter concentrates on (b) and (c). Experience (a) is idiosyncratic and therefore difficult to discuss formally.

In the following, we contrast the use of HF research literature for explanation of phenomena involving technology with the use of that literature for the development of design guidelines. The reason for this is because we wish to cast new light on what guidelines are, their special nature, and the significant processes that must be performed to develop them.

A design guideline is a behavioral principle (derived ideally from experimental studies) that attempts to describe how a physical characteristic of an SEP should be designed to maximize operator performance with that equipment, considering of course the operator's health and safety. Examples of what are considered design guidelines from the software field are to be found in Williges and Williges (1984), Williges, Williges, and Elkerton (1987), and Smith and Mosier (1986).

We have more to say about such guidelines later, but the reader should be aware at the outset that guidelines such as these are useful primarily as a check on the adequacy of an already developed design.

There is considerable variation in the information provided by journals and books. We identify six types that vary on the following dimensions: abstractness/concreteness, qualitative/quantitative, and advisory/mandatory. The six types, starting with the least concrete/qualitative and advisory are: (a) general methodology, (b) attributes, (c) criteria, (d) general guidelines, (e) specific guidelines, and (f) requirements (as in standards).

A general methodology is least concrete, most qualitative, and only advisory. It provides a context for design while not suggesting a particular design technique. It presents an approach to or manner of design. The previous discussion by Rasmussen et al. (1994) is an example of a general methodology.

Attributes and criteria are closely related. An attribute is a quality one wishes the design to possess. For example, the widely used attribute *user friendly* appears appropriate to all equipment. An attribute can be considered a design goal. It is relatively abstract, advisory only, and qualitative.

A criterion takes the attribute and extends it as a measure of personnel performance. Thus, *user friendliness* may be defined by a low error rate, no overt signs of difficulty in the operator, and so on. Neither the attribute nor the criterion suggests how user friendliness, for example, is to be included in the equipment.

A general guideline makes a specific recommendation for design. For example, Rubinstein and Hersh (1984) presented their design recommendations specifically as guidelines (e.g., "Guideline 38. Avoid multiple style modes"; p. 113). This type of guideline is qualitative, although it may be relatively specific. It is advisory only.

A specific guideline is much like the general guideline, but its scope is more delimited. Moreover, it is quantitative. An example might be: The size of the type font must be at least 8 mm and preferably 12 mm for adults with normal eyesight.

A standard is a specific guideline that is no longer merely a recommendation that a wise designer will follow; as a standard, it is mandatory, and violations of the standard must be justified. Because they are mandatory, standards are issued very formally; they are produced by authoritative agencies like the government or institutions specifically established to develop standards (e.g., the American National Standards Institute [ANSI]).

The guideline has certain special features. Rarely does the individual study from which it is derived deal with specific design features (unless of course the study was initiated to answer a question about those features). Consequently, the original study material (data, conclusions) must be changed in several ways: by developing a general principle (something like developing a small scale theory) and applying that principle to some explicit or implied physical feature of a type or class of equipment. All of this presumes transformation processes of greater or lesser extent.

We start with an assumption: If, as we maintain, HF is a design-related science, then its research can or should be utilized by HF personnel to solve their design problems. One purpose of the HF research is to explain behavioral phenomena related to technology. Another purpose, no less important, is to use the results of that research to derive guidelines that aid in the development of equipment. The first function may be considered more scientific than the second, but if only the first function is performed, the research is not HF but psychology.

There are similarities as well as differences in what one does with the HF literature in both of the prior situations. In both cases, deduction (which is a form of transformation) is required; but the transformation is much greater in the development of guidelines because the guideline represents an attempt to cross the behavioral/physical barrier. Use of the research to develop conclusions about human reactions to or use of technology is much closer to the original studies because, to develop these conclusions, no domain crossing or transformation is required. In such studies, we are dealing with behavioral (psychological) phenomena only; the technology serves simply as stimuli to the human subject, who is the focus of the study. The study of a behavioral phenomenon attempts to answer a single question only: How did the phenomenon arise (presumably through some interaction with technology)? The design problem study may have to be applied to multiple and more specific situations because each design problem to which the guideline will ultimately be applied is individual, although the individual problem may be, as suggested previously, only one of a class of design problems.

In many of the studies attempting to explain behavioral phenomena through their interaction with technology, that relationship of the technology to the behavior is minor; in the research that produces the useful guideline, the technology is central to the study. The difference can be seen in two related (hypothetical) studies. In the first, the research seeks to determine the minimum type font size for effective human performance. In a second study, the ability of the elderly to read medication labels is assessed. In the second study, there is an interaction with type font size, but that size is not explicitly extracted as the theme of the study.

In the first case, technological parameters produce the study; in the second case (which is the most common one), it is the behavioral parameters (i.e., the ability of the elderly) that initiate the study. It may be possible to derive technological answers from the second study, but with much greater difficulty. HF research as a whole deals with technology in a trivial way, which HF as science can perhaps accept, but which HF in its application mode cannot. Because of the necessity of relating the behavioral principle to the physical equipment, HF design research faces much more stringent criteria.

Sometimes the transformation is easy. For example, the prescription that a menu should have no more than N levels is derived from studies in which subjects were asked to access files using various menu levels. This transformation of research into guidelines is simple because there is an equivalence between the menu in the research and the menu in the software.

General guidelines are less easily transformed into specific ones because the number of conditions to which they may apply are many. "Use a consistent dialogue style" when applied to software design could have different meanings to different designers. Of course, if there is no explicit technological reference in the study, no transformation of the research into guidelines is possible.

For example, a recent study (Maltz & Shinar, 1999) contrasted the eye movements of younger and older drivers. The technological context was a traffic scene image presented to subjects, whose task was to search for numeric overlays. The visual search of the older adults was characterized by more fixations and shorter saccades. On another task involving photographs of traffic scenes, older subjects allocated a larger percentage of their visual scan time to a small subset of areas in the image. The results show that aging affects visual information processing. Potential applications of the study include training of older drivers with redundant information. This study is characteristic of the majority in the HF literature in which technology is merely a stimulus and the application of the study results to technology is at best remote.

The guidelines with which we are concerned describe primarily the human–machine interface. There are guidelines that refer to the anthropometry of the work station (e.g., a computer terminal) and its physical characteristics (e.g., color of the alphanumerics, size of the type font), but most software guidelines deal with characteristics serving as direct informational stimuli to the operator.

These guidelines are molecular, of course; there are apparently no behavioral guidelines that deal with major system attributes such as complexity or transparency, the maximum amount of information that should be presented, or the conceptual structures underlying navigation through software.

Currently available guidelines have two uses. The first, which is minor, is to assist initial design by being a sort of checklist to remind the designer of aspects he should consider. The second, which is more important, is to evaluate that design after it has been formulated. When we say that the guideline should assist design, we do not mean to suggest that it can or should replace the creative process of finding a design solution that is responsive to a use scenario. It cannot do so because specific guidelines are relatively molecular (e.g., suggesting that a graphic user interface [GUI] should scroll up or down). However, once it is decided that a GUI will be used and there will be scrolling, the guideline can suggest the appropriate direction. Once the design is completed, at least initially, the guideline can be used as a means to evaluate whether there are any equipment features that conflict with the guideline.

To be maximally useful, a design guideline should have the following characteristics:

1. A specific behavioral principle should be related to some physical hardware or software feature;
2. The principle should describe the utilization of the principle in quantitative terms, expressed, for example, as a range (e.g., upper and lower limits);
3. The guideline should describe how the principle should be utilized (under what design circumstances and how it can be applied);
4. The effect the application of the principle will have on human performance if it is utilized, expressed in quantitative terms focused on error probability or a similar metric; and
5. How the principle was derived (from what studies and data).

Needless to say, presently no guideline has these characteristics.

There are two questions we must ask about guidelines: How adequate is the HF literature as a whole to supply material for such guidelines? How adequate are the guidelines when they are formulated? Literature adequacy is not the same as guideline adequacy, which means that ultimately we have to examine compilations of guidelines. The adequacy of the literature to supply material that can be transformed into guidelines depends on answers to questions such as: How much HF literature is design-relevant? Of that design-relevant literature, what systems and what user populations does it relate to? Do the design-relevant studies supply any potential guidelines at all? How general or specific are they?

EMPIRICAL STUDIES OF HF RESEARCH

This section describes the authors' attempt to examine HF literature adequacy from the standpoint of guidelines. The analysis and evaluation of that literature ought to be a routine and continuing effort of measurement experts. The general methodology consisted of asking various questions about that literature. However, before one can ask these questions, one must be able to identify those studies that are relevant to design. This was done by examining the papers in the HFES annual meeting *Proceedings* from 1990 through 1997. This is of course a limited sample. Why not go further afield to the journal *Human Factors*, to the *IEEE Transactions on Man, Systems, and Cybernetics*, to the *International Journal of Man-Machine Studies*? The rationale for confining ourselves to the *HFES Proceedings* is that the latter journals are much more academic than the *Proceedings*. As such, they would hardly contain many specimens of design-related literature. We went where the possibilities were greatest.

There is of course the question of how one identifies a design-relevant study. Presumably such a study would discuss a methodology related to the construction of some technology output. This is a very broad definition, but we wanted it to be broad because we did not want to miss anything. We were prepared to accept a wide variety of studies as long as they had some (even peripheral) relationship to the development of a physical system. For example, certain studies dealing with the characteristics of children and the elderly were considered design relevant if it were possible to deduce from them characteristics that should be embodied in the design of systems for children and the elderly.

Having defined, however inadequately, the studies we would accept as design relevant, there were certain questions we asked about them:

1. What percentage of all papers presented are design relevant? We determined this by taking the total number of papers in each year's *Proceedings* and dividing this number into the total of design-relevant papers for that year to produce a percentage. The percentage varies from year to year (e.g., 11% in 1990 to 6% in 1997), but averages out over the years to about 9%. What this means in effect is that 90% of the HF literature (as defined by the *Proceedings*, which presumably contained more design-relevant material than the other journals noted previously) do not supply design-relevant material and are in fact nothing more than psychological studies with a remote technology reference. This is a serious charge for a discipline intended to relate behavior to technology. Cynically, one must assume that this pattern will not change because at least half the members of the HFES consider themselves psychologists (Hendrick, 1996).

2. Another question that was asked was: To what type of system do the studies refer? For example, one would expect that a high frequency of papers would deal with computers, computer software, and graphic interfaces, and this is what we found. However, a great variety of other types of systems was noted. Other than confirming the current level of interest in computers, the answer to this question was not very revealing.

3. Another question of interest was the subject population tested or referred to. The categories here were: adult, aged, child, and disabled. In all years, the overwhelming majority of papers (95% or more) dealt with adults; only one or two in each year involved what can be termed *special* populations.

4. The type of study performed was of interest and is presented in Table 5.1. The categories are not mutually exclusive. For example, case studies often included a methodological description; hence, the totals sum to more than 100%. In previous analyses of the literature (Meister, 1997),

TABLE 5.1
Frequency and Percentage of Design-Relevant Papers by Study Type

Study Type	Frequency	Percentage
Experiment	63	27
Survey	23	10
Case study	77	33
Theory	25	11
Method description	67	29
Model description	10	5
Symposium	7	3
User/usability test	8	3

most studies were experimental, but design-relevant studies tend to be almost equally case studies and descriptions of methods.

5. A highly relevant question was whether the individual study presented guideline-relevant material and, if so, how general or specific it was. The great majority of studies yielded no guideline material (80%), a few (15%) contained general guideline material, and only 5% had specific material. General guideline material was qualitative (e.g., labels should be larger for the elderly); specific material was quantitative (e.g., type fonts should be at least 12 mm).

6. We were also interested in the themes with which those papers dealt. In other words, what were these design-relevant papers trying to tell us? Design is so broad a topic that it can include many themes.

Table 5.2 lists the specific themes under more general themes. The percentage relate to the relationship of specific to general themes, not to the total number of design-relevant papers.

Findings

Despite that design application is a major aspect of the HF discipline, it seems to play a minimal role in the research literature. That literature attempts to explain behavioral phenomena in relation to technology without being concerned with the effort to learn how to create that technology.

In view of this, the likelihood of finding guideline-relevant material in the general HF literature is not too promising unless HF design personnel do their own research. There is an inconsistency here. Smith and Mosier (1986) revealed hundreds of guidelines relevant to software development. Where did these come from? It may be that these were derived from non-HF papers, such as those published in CHI *Proceedings*. After all, only a

TABLE 5.2
Frequency and Percentage of Design-Relevant Themes

General Theme	Specific Theme	Frequency	Percentage
Users	(1) User assessment as design and	48	21
	evaluation criteria	10	20
	(2) Model of the user	1	2
	(3) Methods of eliciting information from/		
	about users	7	14
	(4) User characteristics/capabilities	5	10
	(5) Special user classes	16	33
	(6) User requirements/performance/		
	preferences/needs	9	19

(Continued)

TABLE 5.2
(Continued)

General Theme	Specific Theme	Frequency	Percentage
Usability testing		22	9
	(1) Methods	12	55
	(2) Effects/effectiveness of usability testing	3	14
	(3) Variables in usability testing	4	18
	(4) Factors affecting usability testing	3	14
The design process		79	34
	(1) Description of the design process	8	10
	(2) Design analysis	4	5
	(3) Design methodology (general)	25	32
	(4) Design variables (age, keysize, etc.)	6	7
	(5) Organizational relationships with design	3	4
	(6) Comparison of design methods	9	11
	(7) Simulation models in design/test	4	5
	(8) HF personnel and HF design as part of the design process	7	9
	(9) Operator role in design/automation	1	1
	(10) Design models	8	10
	(11) Ergonomics design effectiveness	3	4
	(12) HF design education	1	1
HF guidelines		24	10
	(1) Development/use of guidelines/standards	16	70
	(2) HF database/ways to present it	8	33
Test and evaluation		24	10
	(1) Rapid prototyping	8	33
	(2) T&E methodology/tools	12	50
	(3) T&E results	2	8
	(4) Experimental comparison of alternative designs	1	4
	(5) Problems of measurement (general)	1	4
Computers		22	9
	(1) Computer simulation in design	2	9
	(2) Comparisons: computer vs. hard copy	9	41
	(3) Input devices/methods	4	18
	(4) User interface	4	18
	(5) Effectiveness: computer-aiding, DSS, expert systems	3	14
Design applications		6	2
	(1) HF in architecture	2	33
	(2) Virtual reality in design	2	33
	(3) Cost–benefit analysis of design	1	16
	(4) HF in maintenance design	1	16
Information utilization		5	2
	(1) By operators	2	40
	(2) Elicitation of information from SMEs/users	3	60

TABLE 5.3
Classification Scheme for User Considerations

1.	DATA ORGANIZATION	2.4.5.3	Immediate Commands
1.1	Information Coding	2.4.6	Command Operation
1.1.1	Color Codes	2.4.7	System Lockout
1.1.2	Shape Codes	2.4.8	Special Operations
1.1.3	Blinking Codes	2.5	Formal Query Languages
1.1.4	Brightness Codes	2.6	Restricted Natural Language
1.1.5	Alphanumeric Codes		
1.2	Information Density	3.	USER INPUT DEVICES
1.3	Labeling	3.0	Data-Entry Procedures
1.4	Format	3.1	Selection of Input Device
1.4.1	Prompts	3.2	Keyboards
1.4.2	Tabular Data	3.2.1	Special-Function Keys
1.4.3	Graphics	3.2.2	Cursor Control
1.4.4	Textual Data	3.3	Direct Pointing Controls
1.4.5	Numeric Data	3.4	Continuous Controls
1.4.6	Alphanumeric Data	3.5	Graphics Tablets
1.5	Screen Layout	3.6	Voice Analyzers
2.	DIALOGUE MODES	4.	FEEDBACK AND ERROR MANAGEMENT
2.0	Choice of Dialogue Mode	4.1	Feedback
2.1	Form-Filling	4.1.1	Status Messages
2.1.1	Default Values	4.1.2	Error Messages
2.1.2	Feedback	4.1.3	Hard-Copy Output
2.1.3	Screen Layout	4.2	Error Recovery
2.1.4	Data-Entry Procedures	4.2.1	Immediate User Correction
2.1.5	Cursor Movement	4.2.2	User Correction Procedures
2.2	Computer Inquiry	4.2.3	Metering and Automatic Error Checks
2.3	Menu Selection	4.2.4	Automatic Correction
2.3.1	Order of Options	4.2.5	Stacked Commands
2.3.2	Selection Codes	4.3	User Control
2.3.2.1	Letter Codes	4.4	Help and Documentation
2.3.2.2	Number Codes	4.4.1	Off-Line Documentation
2.3.2.3	Graphic Symbols	4.4.2	On-Line Documentation
2.3.2.4	Mnemonic Codes	4.5	Computer Aids
2.3.3	Menu layout	4.5.1	Debugging Aids
2.3.4	Menu Content	4.5.2	Decision Aids
2.3.5	Control Sequencing	4.5.3	Graphical Input Aids
2.4	Command Languages		
2.4.1	Command Organization	5.	SECURITY AND DISASTER PREVENTION
2.4.2	Command Nomenclature	5.1	Command Cancellation
2.4.2.1	Abbreviations	5.2	Verification of Ambiguous or Destructive Action
2.4.2.2	Argument Formats	5.3	Sequence Control
2.4.2.3	Separators/Terminators	5.4	System Failures
2.4.3	Defaults		
2.4.4	Editor Orientation	6.	MULTIPLE USERS
2.4.5	User Control	6.1	Separating Messages/Inputs
2.4.5.1	Command Stacking	6.2	Separating Work Areas
2.4.5.2	Macros	6.3	Communications Record

From Williges & Williges, 1984, with the permission of the HFES.

fraction of the references in Williges et al. (1987) on software design came from papers published in HFES journals. Our literature sample may have been too restricted; one possibility exists that computer specialists working on HF-relevant problems prefer to publish in journals specialized for computers.

What is remarkable is that, with so little HF literature related to design, there are as many design guidelines as we have. Table 5.3 shows how wide the range of these guidelines can be (for software at least). The point to be emphasized is that whatever guidelines we have come from studies specifically developed to produce an answer to a design-related question (e.g., what is the minimum type font size which can be utilized effectively?). Guidelines will not come from the mass of HF studies that might as well have been published in psychological journals. The implication is that HF research is, on the whole, not achieving one of the major goals for which the HF discipline was developed.

Guidelines for software development cover many areas, as shown by Williges and Williges (1984), reprinted in Williges et al. (1987). Nevertheless, many of these guidelines appear to be hopelessly general, as indicated by the samples provided by Liu (1997).

Table 5.1 examined the source of the design-relevant papers. Although the experiment is the dominant methodology in the general behavioral literature, where design is involved, the experiment is eclipsed by the case study. This may partially explain why an experimentally oriented discipline like HF does little with design because the experiment is not as convenient a method of investigating design as it is for other HF nondesign topics. We hypothesize that because design is cognitive problem solving, it has many more degrees of freedom than are permitted in a formal experimental design. It is more difficult to put a design question into an experimental format except that of a molecular nature (e.g., the size of intervals within scales). Because design activities occur in large organizations, often involving teams of personnel over relatively long periods of time, it is difficult (although not impossible) to represent the developmental process in the constrained experimental framework. For these reasons, HF people who study design make greater use of observation and interview, experience (as shown in the case study), designer opinions (the survey), and descriptions of method that have been developed.

Chapter 4 showed that it is possible to study design nonexperimentally. It is also possible to perform experimental studies on design processes by setting up small-scale test situations with individual designers, as Meister (1971) showed. Both experimental and nonexperimental methods should be used in studying design.

The impression one receives from Table 5.2 is that many of the design-relevant studies are oriented around methodology, which means that they are general and not necessarily factual. These studies are design relevant, but it cannot be said that most of them were initiated by a specific design question. Many of them represent thoughts about design in general.

Design guidelines are, in most cases, simple aphorisms, as shown by Table 5.4. In fact, they may be too simple; in their simplicity, they may be much too general for effective use. The more general the guideline, the less useful it is except perhaps as a reminder. Designers prefer simplistic concepts, of course, as we all do (Lund, 1997). These aphorisms serve as an entrée to design without having to think seriously about what that design entails. Designers may prefer to cut directly to the more concrete design issues.

General guidelines, such as the one suggesting consistency of dialogue style, incorporate behavioral principles based on logic or common sense. It is reasonable to postulate that humans prefer consistency in the stimuli presented to them; nor would it be necessary to run a study to demonstrate that consistent formats produce more effective performance than inconsistent ones.

Nevertheless, beneath the apparent simplicity of the general guideline are submerged concepts and methods that the specialist must unearth if he or she is to use the general guideline meaningfully. Dialogue consistency requires complex analysis. Aphorisms enable design personnel to assure themselves that they have dealt with the underlying complexities without, in fact, having to do the front-end analysis that is implicitly required by the aphorism. The use of the aphorism as a meaningful design guideline merely encourages design personnel to think simplistically about the design problem at least regarding behavioral factors.

This tendency may be encouraged by the designer's need to simplify complex problems by decomposition, and thus to control the problem. One can draw interesting but speculative inferences about the designer's need to master the problem; this is the motive for decomposition. Designers may feel that behavioral specialists complicate their problems by introducing factors like the human–technology relationship, which designers have difficulty understanding. In contrast, some specialists find this relationship inherently complex and try to simplify it by analysis. The designer simplifies though the use of experience and the aphoristic guideline.

HF specialists have had a reputation from the beginning of the discipline of wanting to study a design problem before providing answers. They may have had the feeling (undoubtedly justified in many cases) that the problem as the designer conceptualized it was more complex (at least from the operator's standpoint) than the designer realized. The front-end analysis inevitably complicated the designer's task, and the designer relishes simplicity.

TABLE 5.4
Samples of Interface Guidelines

Samples of general design guidelines
- Strive for consistency.
- Enable knowledgeable frequent users to use shortcuts.
- Provide information feedback.
- Organize sequence of actions into groups.
- Offer simple error handling mechanisms.
- Allow easy reversal of actions.
- Enable users to be in control of the system.
- Reduce short-term memory load.

Samples of data display guidelines
- Left justify columns of alphabetic data to allow rapid scanning.
- Label each page to show its relation to other pages.
- Maintain consistent format from one display to another.
- Display data in directly usable forms.
- Use short, simple sentences.
- Use affirmative rather than negative statements.
- Provide an informative header or title for every display.
- When blink coding is used, the blink rate should be 2 to 5 Hz.

Samples of screen design guidelines
- Make appropriate use of abbreviations.
- Avoid unnecessary details.
- Use concise wording.
- Use familiar data formats.
- Use tabular formats with column headings.
- Arrange related items as groups.
- Use highlighting to attract user attention to certain elements.
- Present information in a proper sequence.

Samples of color usage guidelines
- Avoid pure blue for text, thin lines, and small shapes.
- Avoid red and green in the periphery of large-scale displays.
- Not all colors are equally discernible.
- Do not overuse colors.
- Use similar colors to convey similar meanings.
- Use a common background color to group related elements.
- Use brightness and saturation to draw viewer attention.
- For color-deficient viewers, avoid single-color distinctions.

Samples of error message guidelines
- Be as specific and precise as possible.
- Be positive: Avoid condemnation.
- Be constructive: Tell user what needs to be done.
- Be consistent in grammar, terminology, and abbreviations.
- Use user-centered phrasing.
- Use consistent display format.
- Test the usability of error messages.
- Try to reduce or eliminate the need for error messages.

Design-Related Research Themes

The major themes in Table 5.2 are: users, usability testing, design process, HF guidelines, test and evaluation (T&E), and computers. Minor themes are design applications and information utilization. The major contributor to the user theme is the special user class because of the small but recurring series of papers on the aged. Usability testing is preoccupied with method; descriptions of the design process, which involve the great majority of papers, are heavily loaded on methodology. The 24 papers describing guidelines are relatively homogenous. There are only two subthemes: the development/use of guidelines and standards, and the HF database and ways to present it. Papers focusing on T&E emphasize rapid prototyping and measurement (i.e., comparisons of computerized techniques with other presentation methods). There are a few occasional papers on special design applications like virtual reality, and a small number of studies on information utilization by operators as a preliminary to design and methods of eliciting information from SMEs and users; these could also fall into the user category.

The preoccupation with users reflects a general concern of the HF literature with the behavioral response to technological stimuli. With general commercial systems, this point of view is very pragmatic because the equipment must be sold. In earlier days, the user was much less considered; the designer asked the customer what his or her design problem was and then proceeded to solve it. Why the change from designer-oriented to user-oriented design occurred (and to what extent) is not clear, although it is very possible that over time HF efforts had a part in this. The computer revolution has obviously made it easier to include the user in the design process, but that is probably not the complete explanation.

The concern for usability tests accepts the proposition that initial design will be less than completely adequate, but that the usability tests and prototypes presented to or tested by the user will discover these inadequacies, which can then be remedied.

Another factor pushing usability and prototype tests is economics. These two methods save time, if not money, because fewer analyses need to be done to determine the more minute factors that could affect operator/user behavior.

Another major preoccupation of design-related HF research is the nature of the design process, which is a welcome development because it is an attempt to understand what HF specialists do and the context in which they work. Many of the general guidelines we can identify herein are of this nature.

Linked to the design process is a tendency to develop computerized HF tools like computerized databases and toolkits. HF researchers are as fascinated as everyone else with electronic presentation devices. There is some feeling that computerization can assist design, and of course it can (e.g., by allowing designers to make three-dimensional representations of their design and rotating these to permit various perspectives). Much of this computer-assisted design is performed by engineers and not for HF purposes, but computerization also assists in prototyping and usability tests in which HF is or should be much involved. From the HF standpoint, the great advantage computers will present in the future will be their ability to manipulate human performance simulation models. Here, too, however, it will be the HF specialist who will create the model and input the data. The computer will then put the model through its paces, changing variables, modifying data inputs, and presenting model outputs to the specialist. When the use of human performance models in design becomes a routine event, the computer can be said to have come into its own as an HF tool. Behavioral inputs such as the task analysis and the mission scenario, which are embedded in these models, will then automatically become required inputs to design.

If design guidelines are derived from the literature, one must go beyond the guidelines to ask whether HF literature, in general, is useful to practitioners. The answer is to be found in Chapter 4 on design practice, in which we asked that question as part of our survey of those practices. The impression we get is that HF literature has only limited value.

One reason that HF research literature has only limited value is that, as a whole, it is undigested—simply, many papers on many topics. It can only become useful after it is analyzed and combined, its salient elements emphasized, and its quantitative values extracted—the way mother birds predigest food for their babies. Then it will be meaningful and useful to the specialist. As we have pointed out, however, this demands a deliberate, sustained effort to transform and refine that literature just as one refines sugar or raw diamonds.

STANDARDS

Behavioral design standards like MIL-STD 1472F (1999), which all of us who have worked on military system development projects are familiar with, are descriptions of data (e.g., anthropometric) and design practices that are accepted (considered beyond serious challenge). The history of standard development, if one were written, would be fascinating because it involves progressive refinement and considerable infusion of personal beliefs of behavioral designers. For example, MIL-STD 1472, which has just been revised, was derived by a process of refinement from MIL STD 803, an Air Force document of the early 1950s. It then became 803A and progressively evolved into 1472—a DOD document. The refinement included a great deal of discussion; the senior author recalls attending meetings involving dozens of what were then called *human engineers*, at which discussions of what to include in the standard and how it should be phrased were quite vehement. Over the years, questions of what is important in standards have been raised and answered, and raised again. Certainly, opinions of those with expertise and reputation in the standards field (and it is most certainly a field of scholarly concentration and expertise) have greatly affected what is finally published as a government document. The significant aspect of the behavioral standard is that it is mandatory; it does not have the force of law, but it has compulsive force behind it when it is written into contracts as a governing document. Recently, however, in the United States at any rate, there have been attempts to reduce the control of design standards over military systems.

Because of this, certain questions deserve investigation:

1. Where do standards come from (i.e., research sources, expertise)? How much objective research data and how much expertise enter into them?

2. What design questions or topics does the standard deal with (because behavioral design has many aspects)? Which ones are included in the standard determine in part its potency?

3. Are standards tested to determine that they are effective, correct, or valid? Is a standard accepted on the basis of a consensus of those most knowledgeable about a topic (such as ladders)?

4. How comprehensive are behavioral standards (this raises the further problem of what one means by *comprehensive*)?

5. How much information and detail should be included with any prescriptive statement in the standard? For example, should one know the research citations that were considered in writing the statement? Should contextual detail (e.g., conditions under which the statement is particularly important) be included in the standard? (If so, this makes the standard like the design guideline because the contextual material suggests that there may be situations in which the standard should not be applied.)

6. How mandatory is the standard? Under what circumstances must (rather than should) it be applied? (Within the document, there are topic areas with specifications that only apply to that topic area. All sections of a standard are seldom applicable to a specific design.)

7. Can the standard be used to guide design or is the standard only a means to verify a level of design adequacy? If the standard is mandatory, it assumes the character of a warranty, which assures adequacy. Can a behavioral standard do this?

8. What are the anticipated effects of the standard?

Standards are important not only because they control design (more or less), but also because they summarize what experts believe is known about behavioral design (not suspected, not hypothesized, not speculated about). Therefore, the standard can be considered a summary of present HF knowledge about various design aspects, and what is not in the standard can be considered a research problem (what we need to find out).

It is also worth reviewing Smith (1988) on design standards and guidelines. Smith made a distinction among design standards, design guidelines, design rules, and design algorithms. As we said earlier, standards are more or less mandatory, guidelines are recommendations for a particular system, and algorithms are computer software for implementing design rules. Only the first two need concern us. Standards are published in official documents, whereas guidelines may be found anywhere. MIL STD 1472 is published by the Department of Defense; Smith and Mosier (1986) is published by the U.S. Air Force, but has no official standing unless called out in a design specification.

Outside of the legalistic aspect of standards, the difference between standards and guidelines is the amount of contextual material each provides, which is/can be somewhat greater in guidelines. Designers dislike standards because they are restrictive; they are more accepting of guidelines because they are tutorial, and the designer has the option of using or ignoring them. Deviation from standards requires formal official exception, which guidelines do not. Smith (1988) made the point that hardware standards are based largely on physiology (e.g., the intervals in display scales are based on what the eye can resolve); software standards are based on cognitive capabilities, about which much less is known. Because of this, it is likely that there are fewer software standards than there are hardware ones, although in the past 10 to 20 years, there has been increasing interest in software.

The more general the standard or guideline, the more tailoring it requires (application to the specific design problem). The designer needs guidance in making the application, which is why the more contextual material provided, the better.

Smith noted the conflict between the need to phrase guidelines generally for broad application and the designer's need for specific rules. This means that the guidelines must be translated or converted into specific rules. Translation is a familiar process to designers because system development in general requires a series of translations — from operational needs to functional requirements, from requirements into design specifics, and from specifications into component designs. In the case of behavioral guidelines, the questions are: Who makes the translation? How is it accomplished? The first question is easy from our standpoint: It is the job of the behavioral specialist to make this translation. The second question, however, remains an unanswered one.

Smith suggested that when design rules are agreed on, they might be given weights, indicating relative importance. This would be useful in design trade-offs and later design evaluations. He went on to say that guidelines cannot entirely take the place of expert design consultants because the expert will know more about a topic than can be expressed in the guidelines, more particularly what questions to ask as well as the answers to the questions; and how to adapt a general guideline to the specifics of the individual design.

Guidelines do not replace other behavioral tools like task analysis and indeed may require a more detailed task analysis if the former are to be utilized properly. The use of guidelines does not save time or work because it is a step added onto the basic design process.

One of the questions that arises about design standards and guidelines is the extent to which each is based on research or simply codifications of expertise and accepted practice. Smith (1988) suggested the latter (at least with regard to software material), but it must be recalled that his judgment may represent the state of the art as of 10 or more years ago.

We asked Jerry Chaikin, who is a recognized expert on standards and a member of governmental bodies that control these, for his opinion on the matter with particular reference to MIL STD 1472, which has been in existence at least 40 years. Part of his response (Chaikin, 1999) follows:

The first question . . . —to what extent are present day human factors standards based on experimental data?—isn't easy to answer. I'd only be guessing. Using acoustical standards as an example, provisions on maximum steady-state noise limits, detectability, speech interference, and impulse noise are largely quantitative and experimental data-based; provisions on drive-by noise are regulation-based, and helicopter noise, reflecting state-of-the-art-based compromises, might be characterized as consensus- or SME-based. I'd suspect that provisions (in human engineering standards) on anthropometry, environment, and materials handling would be experimental data-based, while most provisions on designing for the maintainer, for example, would not.

I agree with your contention that very little of the standards developed since 1990 is concrete and quantitative. First a tremendous quantity of software ergonomics standards (style guides, UCI, etc.) have flooded the human factors standards arena and, in most instances, seem to be guidelines based on logic, practices that have become de-facto standards, organizational practices, lessons learned, and other convenient foundations necessarily characteristic of a field that is advancing at warp speed. The writers of these standards—guidelines—likely recognize that innovation in such an environment shouldn't be stifled by standards and tend to cover a lot of material via principles and tutorials that provide wide latitude for designers to meet their performance aims. A second reason for my agreeing with your contention is the fact that a lot of commercial and international standards are being developed which tend to satisfy the least common denominator of its participants, thereby diluting hard numbers that might otherwise be based on science (of some sort).

Where does material for 1472 come from?

- Experimental data (more than you might think, but less than it should)
- Lessons learned (usually repeatedly)
- Arbitrary decisions (Army's olive drab; Navy's battleship gray, Air Force strata blue)
- Compliance with laws, regulations (no choice here)
- Harmonization with other recognized standards (what were their sources?)
- Common sense (accessibility to items on the basis of expected frequency of maintenance)
- Observed population stereotypes (light switch up for ON in the US)
- Consensus consideration of user inputs (on what are the user inputs based?)
- SME best estimates that survived the test of time (design of controls?)

In summary, a lot of experimentally based, quantitative provisions appear in various human engineering design standards—probably developed before 1990 and in the hardware and environmental areas. A lot of the qualitatively expressed guidelines and preferred practices emerging since that time are, I believe, in the software ergonomics or UCI arenas where flexibility is needed and in commercially steered standards that seem to strive toward reasonable accommodation.

6

Software Design

We start with a single basic question: What is distinctive about software design? How does it differ from any other design? This question assumes that one can, in fact, distinguish software design from any other type of design considering the ubiquity of the computer these days. The difference, if one exists, may be only a historical one: How is today's design process different from that of, say, 1950? Yet as we saw in our survey sample described in chapter 4, many professionals define themselves as specialists in software design, as opposed to a different (hardware) design, however ambiguous that other type of design is. This suggests that they feel a difference, whatever that difference is.

This specialization justifies a chapter on software design. However, to answer that question, it must be decomposed into a set of more molecular subquestions:

1. What are the characteristics of software design and how does it differ from hardware design?
2. How effective is software design and what are the major problems to be overcome in that design?
3. What kind of HF research has been performed to overcome those problems?
4. For the HF design specialist, software design must almost be equated with the operator/user and the design of the HCI. How does the user and his or her requirements affect software design?

To anticipate the answer to the first question, the general developmental process in software is the same as it has always been for hardware, but there are significant differences in the way in which that general process is implemented.

Those differences involve our belief that software, as it is presented to the operator/user, speaks more directly to the latter because it engages the cognitive function more than is/was the case in non- (or less) computerized systems.

The operator/user has always been involved in design, but usually more covertly, less prominently than in today's design. Cognition equals information-processing, and consequently information-processing theory is a major organizing principle for control of the computerized equipment. Designing the internal components of a computerized system, as in a nuclear power plant (NPP; e.g., making boilers develop steam water to flow in specific directions at certain times or states) does not require information processing (only in the most fanciful way can one say that the boiler receives information, which tells it when to vent steam). However, as soon as the designer wishes to use an external mechanism (i.e., the operator) to control that process, the designer must deal with information.

Because control of the computer requires information and information can only be utilized by the operator/user (ignoring claims by AI theorists), there is much greater concern for the characteristics of the operator, as exemplified by the proliferation of models, such as GOMS, describing the operator/user.

On a more molecular level, human–computer interface (HCI) design makes use of devices not found in noncomputerized systems. Software design requires the writing of programming code; this is peculiar to software. Beyond that, there are special software tools: menus, icons, the mouse and the keyboard, and "widgets," among many others.

Therefore, it is possible to see some characteristics that absolutely distinguish software from anything else (e.g., code), other characteristics that represent an increase in emphasis (e.g., concern for the operator/user), and still other characteristics, such as the general development process, which remains essentially unchanged except for its context.

THE SOFTWARE DESIGN PROCESS

General Outline

We make use of Fig. 6.1, adapted with modifications from Liu (1997), to list the stages in software development and the major HF inputs to these stages. The remainder of this section follows these stages.

The general developmental process depicted in Fig. 6.1 is essentially the same as that ordinarily employed in design and described in previous chapters. The terminology may have been changed, but the reader should not be confused by this. (Lacking an authoritative taxonomy, authors are free to play word games.) For example, Requirements and Task Analysis are part of what is ordinarily called *preliminary design*. Design Generation and Prototype Testing are part of *detailed design*. Human–machine interfaces have always been developed, but in presoftware days were called *control panels* or *workstations*.

Other differences exist. Prior to the advent of computer systems, prototype testing was almost unheard of, although, when serious questions about design alternatives arose, which could not be answered except by testing, mockups in Styrofoam, cardboard, or wood were constructed and subjects were tested with these. Mockup tests were not as common as prototype tests are today simply because the physical mockup was time-consuming and expensive to construct.

The interface development represented in Fig. 6.1 was not ordinarily designated as a special developmental phase. Much less attention was paid to securing information about user activities, although subject matter experts (SMEs) were consulted, and the effort to secure design-relevant information was, as it is now, a significant aspect of development.

Software design, as with all design, is a process of decomposing major physical and behavioral functions into more manageable components and tasks. System requirements must be analyzed and their effects on design possibilities considered. One or more designs must be generated and evaluated in terms of whatever criteria already exist or are to be developed. As indicated in previous chapters, we distinguish between evaluation and testing, the latter involving measurement of subject performance.

Although interface development is accorded a later position in the developmental sequence depicted in Fig. 6.1, we see it as beginning in parallel with functional design (internal components) and being a critical task in each of the subsequent stages. This may be idealistic, but it is also logical. Although interface details must wait on the development of internal components, they must not be far behind because the ability of operators to recognize equipment status and to diagnose impending malfunctions of internal components should be a consideration in the functional design.

Figure 6.1 suggests only some of the HF design methods that can be applied to implement software design.

One major engineering input to system development has not been indicated because it is of little HF interest; this is the programming of software code, which proceeds in parallel with design generation, prototyping, and iterative redesign.

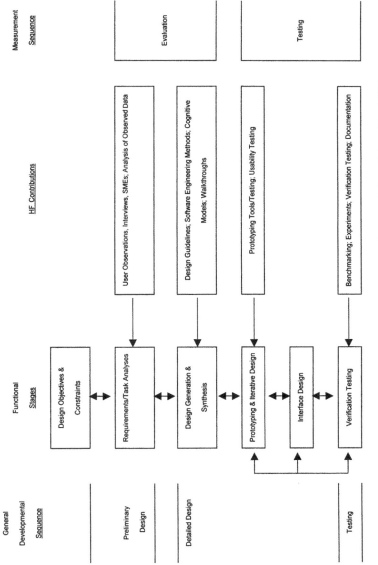

FIG. 6.1. Computer system design and HF contributions. (Adapted from Liu, 1997.)

157

Like all design, software development is goal directed primarily to achieve a certain functionality (competence) to implement functions and tasks demanded of the computerized system. In the next section, dealing with design objectives and constraints, we shall encounter other design goals, but to the practicing engineer the objective of functionality is primary.

The generation of design alternatives involves the development of hypotheses about methods of implementing functions; these are accepted or rejected after evaluation, which involves the application of criteria, such as feasibility, cost, reliability, and so on.

Design Objectives and Constraints

Design objectives and constraints are external factors imposed by the customer, the state of technology, the availability of money and time, and so on. This is not a stage of equipment design; it is the preliminary background for all subsequent development. Part of this background includes what the designer and HF specialist bring to the design process in terms of knowledge, skills, and experience.

Design objectives are abstract goals, theoretical system attributes, which presumably guide design, but only in a very general way. Design objectives differ from design requirements, which are, it is hoped, very specific (e.g., the aircraft must fly at mach 3). Design objectives must usually be inferred by design personnel. It is almost unheard of for such design goals to be described in design requirements documents because they do not have a direct, guiding influence on design.

Design objectives apply to all systems, because of their generality and because they are part of the engineer's disciplinary ethic: Successful systems are those that function adequately, are easy to use, come in under cost and on time, and so on.

Software designers share these design goals with all other designers. These goals include: effectiveness, efficiency, comfort, and safety. The most important goal is effectiveness, which is merely another term for functionality. If the system cannot do what it is designed to do, it is worthless. There are different degrees of functionality; when the designer considers in the Design Generation phase alternative ways to implement requirements, he or she has in mind the greatest degree of functionality that can be secured, considering design constraints.

All other goals are subordinate to functionality, but they serve as constraints on and modifiers of that functionality. For example, no design alternative that would be unsafe for the operator/user would be accepted if the hazard were known. Efficiency encapsulates the notion of degrees of functionality so that, given various ways of achieving performance of a function, some require fewer components, less expenditure of energy, and so on. Therefore, these are to be preferred. Comfort is of lesser importance and certainly has no role in the design of internal components; it may have value in the design of the HCI because only the operator/user can experience comfort or its reverse — discomfort or workload.

As goals, these attributes have little immediate significance; however, when they are used as criteria to evaluate design adequacy, they have much greater importance. They need not be specified formally, however, because it is assumed that they are part of the knowledge inventory all skilled designers possess.

Theorists specializing in the HCI have developed goals more specific to the interface. Norman (1983), Rasmussen et al. (1994), and Vicente and Rasmussen (1992) advised us that the content of the interface should abstract the critical features of the system (and of the world in which the system is embedded [our addition]). This means that the interface becomes a model of these features. This is almost a tautology because the HCI is supposed to model the system and world. However, this objective reminds the interface designer that he or she must study the target system and the world to be represented in the HCI to ensure that all essential features of that system and that world are represented in the HCI. Of course, the question is: What is essential? We see in a later discussion of HCI development that if the system and its world are more than extremely simple, the HCI cannot contain their every characteristic or display their every operation — hence the need to study the system and the world prior to design. A major part of the decision making involved in HCI development is the determination of what to include in the HCI and what to ignore.

The term *the world* in which the system is embedded needs some explanation. Some (perhaps most) systems are used to control processes, as in a nuclear power plant. Some systems, like radar or sonar, are used to observe the world around them. When the system involves the latter, attention must be paid to how that world is depicted by the HCI.

Other theorists tell us that the form of the interface should be consistent with human perceptual and cognitive characteristics (Carroll, Mack, & Kellogg, 1988; Tullis, 1988; Wickens, 1992). This objective may be subsumed under the prior comfort goal. It suggests that no HCI design making excessive demands on human capabilities will be acceptable. (But one must know what is excessive.) This objective acts as a modifier of the functionality goal and can be transformed into an evaluation criterion. At the same time, the implementation of this objective depends on available knowledge secured from research about those human capabilities. Given such knowledge, this goal can be transformed into a measure of HCI adequacy from the operator/user standpoint.

Another objective is that interface manipulation should be compatible with human response tendencies (Shneiderman, 1983). This is merely an expansion of the preceding objectives.

More meaningfully, Williges et al. (1987) developed seven design principles that are not so much design goals as factors to be kept in mind by the HCI designer. These principles are compatibility, consistency, memory, structure, feedback, workload, and individualization. A well-designed HCI should (a) conform to population stereotypes in presenting software symbols and should minimize the need for information recoding or interpretation (compatibility principle); (b) maintain the same style of interaction throughout operations (consistency); (c) minimize memory demands (memory); (d) help operators/users to understand the structure of the system (structure); (e) provide feedback and error-correction aids (feedback); (f) keep user mental workload within acceptable limits (workload); and (g) accommodate individual differences (individualization).

These are useful things for the HCI designer to keep in mind, but there is more to these principles than that. Each of the principles requires for its implementation a set of knowledges that may or may not be available to the designer (more the latter). For example: What are population stereotypes in information processing? How does one measure memory demands? How do users conceptualize system structure? What are acceptable levels of mental workload? How does one recognize the factors that determine these? What individual differences in HCI operation exist? How can they be accommodated?

The real value of the Williges et al. design principles is that they point out where research is urgently needed to aid the designer. They can be used as a conceptual structure to organize that research because no one with even a slight acquaintance with the HF literature would consider that the knowledge needed to answer these questions presently exists.

In Chapter 5, we differentiated design-relevant research directed at the human response to technology from research directed at understanding how behavioral inputs could be incorporated in design. We implied that the latter type of research was more desirable than the former. This point must be clarified.

We now discover a third type of HF research: research directed at determining human capability as that which can be related to technology. Just as we want to know the anthropometry of the human or what strength the 50th-percentile male can exert with his arms, so we want to know the strengths and weaknesses of the average operator is receiving, interpreting, and integrating stimuli (particularly of the cognitive type). The focus is on information and cognition because that is the dimension with which software systems primarily deal.

Shneiderman (1992) had his own design goals in terms of four classes: (a) functionality; (b) reliability, availability, security, data integrity; (c) standardization of features and portability across multiple applications; and (d) development within schedule and budget. Shneiderman's goals are more engineering-oriented than are those of Williges et al., which are essentially behavioral. From that standpoint, Shneiderman's goals require little behavioral research and suggest little of behavioral research significance.

Rasmussen et al. (1994) emphasized giving greater attention to representing abstract, functional relationships involved in the design of the HCI. The traditional structural design strategy, they say, is to transform larger into smaller entities; this represents and analyzes the system by decomposing it into its structural elements. This may be satisfactory as far as it goes, but to Rasmussen and his colleagues the approach is not adequate to model higher level, goal-oriented behaviors. The functional methodology of Rasmussen et al. complements the traditional one by looking at the entire system, representing it by abstraction to an appropriate functional level, and separating the relevant functional relationships. The methodology is described in brief by the term *abstraction hierarchy* (Rasmussen et al., 1994), which we encountered previously.

Liu (1997) pointed out that, in different application domains, operators/users have different relationships with the system. In supervisory control applications, such as air traffic control, operators need to exert immediate, authoritative control over the system and the world (Sheridan, 1984). For applications such as highway traffic administration, operators/users influence the system by issuing advice, warnings, and so on (Murray & Liu, 1995a, 1995b). In reconnaissance applications (e.g., radar, sonar), operators use their system to understand the external environment (the world). Other applications might involve relating and understanding items of knowledge (e.g., an intelligence analysis or stock market system). All of this suggests that the HCI designer must understand the functions and tasks to be performed by the user because HCI design is subtly influenced by the latter's functions and tasks.

The overall goal is functionality; HCI design should assist the accomplishment of designated tasks and missions. Everything else pales in comparison with this requirement. The question is, of course, whether functionality alone is sufficient. Many engineers become so immersed in the functionality goal that all other considerations, including those which the HF design specialist considers important, are ignored. This tendency is aided by the inability to specify quantitatively the relative value of behavioral goals in relation to functionality; this is a serious weakness in HF design methodology.

One must ask whether, from a practical standpoint, these goals (other than functionality, of course) have any strong influence on design. We suspect not in initial design, but their influence can be strengthened if they are transformed into evaluation criteria and impact redesign. Their effect can be enhanced if, as criteria, they are buttressed by quantitative design-relevant HF research.

Requirements and Task Analysis

This phase of computer system development corresponds to the traditional preliminary design phase. Requirements analysis is the analysis of what the system is asked to do; task analysis describes what the operator is asked to do. The structural decomposition strategy is generally followed. The overall computer mission is deconstructed into those functions that must be performed by internal components and those that must be performed by the operator.

Hence, there are, or should be, two parallel sets of analytic activity — one dealing with the system as it controls external world processes and another designed to determine what the operator must do. Outside of the dedicated software (embedded software used to control internal mechanical and electrical processes), computer software is designed to permit the operator's interaction with the system and the world — hence the importance of data concerning user capabilities.

In both cases, the analytic techniques center around questions to be answered and the implications of the answers. We discuss the kind of questions that are asked in the section on design generation. However, the questions generally deal with the nature and dynamics of the world and the computer system used to control that world. The operator is part of that system, and so extensive efforts must be made to describe precisely what the operator does in achieving that control.

Such questions generate further questions relative to operator capability. It is not enough to determine what operators must do; once this is learned, it is necessary to determine whether any special operator capabilities or disabilities are required by or implied by these activities. Information about general human capabilities cannot be gained by testing operators of a specific system, but should be provided by HF research.

Viewed in this light, it is remarkable that so much more is known about the physical properties of components than is known about the behavioral properties of the operator. No one would design or purchase a boiler without knowledge of the steam pressure the boiler structure is capable of withstanding. However, beyond certain fundamental thresholds of sight, sound, and motor abilities, everything else about the human operator is simply assumed to be adequate by the specialist and engineer. For example, we would obviously not put a mental retardate in charge of the system for monitoring a chemical processing plant, but beyond that simple reservation we know practically nothing about the relationship between equipment complexity and intelligence.

It is assumed that the new computer system will automate certain functions and processes formerly performed manually, or that it is an upgrade or modification of a predecessor system that also had operator personnel. In either case, it is necessary to learn more about what the operator of the new system is supposed to do. This is particularly the case when the new system will automate previously manual operations.

In this situation, the HF design specialist will perform his or her own naturalistic research, which is parallel to that which HF researchers do experimentally. The specialist will (a) observe routine activities; (b) ask for demonstrations by the operator of how he or she performs tasks (both of these video recorded for later analysis); (c) interview operators and SMEs; (d) use questionnaires to collect information; (e) meet with focus groups; and so on. We consider these activities to be design research, but performed with fewer experimental controls and for highly specific purposes. Earlier design did all these things also, but not to the same extent, nor did it recognize the importance of learning about user activities.

The information sought by these techniques consists of the following questions: What does the operator do and how does he or she do it? What is the function of the activity? What factors affect its performance? What information does the operator seek as part of this activity? How does he or she process that information and what does he or she do with it? What internal and external factors affect the information-utilization process?

The information gained in this way serves as inputs to a formal task analysis (TA) and a variation of that TA—cognitive task analysis (CTA). These have been previously discussed.

The goal of the activity in this phase is to develop a set of relationships between the decomposed functions and tasks. These relationships can be expressed in the form of a sequence, hierarchy, or network of task events. Each of these is a way to indicate a logic on the basis of which software code can eventually be written. Programming, as an engineering activity, is outside our purview. Hence, it is not discussed here.

Liu (1997) suggested that models of cognitive performance can be extremely important in organizing how the task components can be interrelated. Presumably, if data about user activities can be input into any of these models, the underlying logic of the model can be transformed into software logic. A more detailed description of these models is postponed to a later section.

There may be several outputs of the TA and the data gathered by observation, interview, and so on. There is the verbal CTA, which lists in sequence the stimuli, information-processing, and response tasks to be performed. As indicated previously, the CTA can be expressed graphically as flow or network diagrams, or as scenarios, either in paper or electronic form.

This permits the application of more or less formal evaluation procedures. Such evaluations involve criteria that may be derived from system development goals (see preceding section) or may be sought in design guidelines.

The development of evaluative criteria should be a formal effort performed in parallel with the CTA. It is not sufficient for the HF specialist or software designer to rely on intuitive processes ("I know what is wrong when I see it"). If the goals described in the preceding section have any meaning or value at all, it is in their transformation into evaluative criteria. These criteria already exist as goals, which are only abstractions; they must be tailored (made more concrete) to the specifics of the new design.

Design guidelines can also be used as evaluative standards. These cannot be used for initial design, but they can correct any detailed design inadequacies. Of course, the analysis must proceed to the stage of preliminary design before guidelines can be applied; they cannot be used to evaluate analytic outputs (TA, CTA, or any TA derivatives).

Preliminary software and/or interface designs may be developed at this stage. In the traditional developmental process, the distinction between one stage and another is usually blurred. The preliminary design stage in traditional system development often included preliminary designs for evaluation; they may do so, also, for computer system development.

The TA and CTA used to model the system and the real world cannot be complete, as pointed out previously, because the model cannot contain all the aspects of the real world; consequently, it may be off kilter to some extent. The model may even be deliberately biased by preconceptions. For example, in their book *Decision Support Systems*, Keen and Scott Morton (1978) pointed to the necessity of determining the approach the work force will use in carrying out their tasks. They gave an example of an investment firm whose head wanted a particular approach (investment strategy) followed to the exclusion of all others. The DSS program and its support data were so designed that only that strategy and no other could be used. Outside of the implied question of for whom the system should be designed—owners or system operators/users—the more obvious question is whether the selected computer approach modeled the real world correctly or as efficiently as it should have.

The adequacy of the computer software model (its functionality) cannot be completely determined except by verification testing (OST). Heuristic evaluations, prototyping, and usability testing do not answer the question of adequacy because all such preliminary evaluations and tests do not place the system into the actual or simulated operational environment, so that a comparison between the system and real-world functioning can be made.

Design Generation and Synthesis

One can think of Requirements and Task Analysis as the stage in which the design team determines what must be designed. The Design Generation stage, into which it melds unconsciously, is the stage at which the team decides how the system is to be designed. Alternately, one can think of the analysis stage as determining what the problem is and the Design Generation as the stage in which the team solves the problem.

Refer to Chapter 4, in which the team considered various ways in which an ATM could be designed. The functions to be performed by the user with the ATM had been specified; now the question to be answered was: What is the best, most efficient way to input the user's PIN number, of receiving feedback about the status of the account, and so on?

The emphasis on software in design terminology misses the point. Actual design decides how to implement certain functions and tasks. The software, in terms of programming code, follows whatever problem solution is decided on as the best.

It was pointed out previously that specification of functions and tasks does not inevitably lead to one and only one design solution. There may be a number of ways to implement a function and task; that is what was meant by degrees of functionality. Each design proposed may well solve the problem, but only one solution will be most functional (i.e., most efficient, etc.; see the design goals in the earlier part of this section).

Design generation is the development of hypotheses to answer the question raised by the specification of functions and tasks: How should these be performed? Each design hypothesis is evaluated for its functionality; that is why the need to develop clear, objective evaluation criteria has been emphasized. Design generation (once a solution is proposed) involves two sets of comparisons: (a) a comparison of each potential design solution with external evaluation criteria embodied in such instruments as design guidelines, and (b) a comparison among alternative design solutions in terms of those criteria.

One can think of design generation and its evaluation as two sequential phases, but actually these two processes are usually interwoven. The rejection of one potential design solution requires the development of another to replace it. At the beginning of the design-generation activity, the members of the design team may develop a number of design possibilities and only gradually whittle them down.

Although it takes place in a team context, the development of design solutions is a very personal, idiosyncratic effort, but with the other members of the design team analyzing, commenting on, and suggesting modifications to the original concept—a sort of Greek chorus.

There is a need to understand the process of design generation: Being a creative function it appears mysterious, and because it is mysterious, it lacks the control one seeks to exercise over the process.

The techniques developed to help solve the design problem (e.g., network diagrams) are created at least in part to constrain the less than fully rational aspects of the design process, and to force a more systematic approach on designers. The irrationality stems from the creative, intuitive, and idiosyncratic nature of designers. We do not suggest that designers perform randomly, but their thinking can veer into byways. One need only read the excerpts of design deliberations in Chapter 4 to recognize that such techniques help design be more systematic than it might otherwise be. Control is exercised by development of formal techniques to force the designer to follow their logical requirements. A task analysis, for example, may force the designer who reads it to follow the mission sequence that is reflected in the TA. Flow charts, network diagrams, scenarios, storyboards, documentation, and so on are all control as well as analytic mechanisms.

Even informal processes can serve as control mechanisms, although the preference is for formal ones. When, as in prototyping, we involve the user in the process of evaluating design, we give up control (in part) to the user. Therefore, the users become, without their knowing it, control mechanisms. Experimental comparison of alternative design options is another form of control, as is the evaluation of intermediate and final design products.

We referred in the past to something called Design Rationale (DR; Moran & Carroll, 1996), to which much of the following is indebted. DR is an attempt to control and focus the many ways in which software can be developed. The control is produced by forcing designers into a formal manner of conceptualizing the process of arriving at a design solution.

Control is required by the special characteristics of design. The most obvious characteristic of the design process is its inherent complexity. This forces the designer to decompose the many aspects of the design problem into more manageable subdesigns. However, decomposition is problematic (i.e., there are different ways of decomposing). Decomposition requires recomposition or synthesis, in which the individual subdesigns must be combined to form an overall configuration. Synthesis may be almost as demanding as decomposition. Part of the reason for DR or any formal methodology is that, as Moran and Carroll (1996) suggested, the needs being addressed in design are sometimes poorly specified — vaguely stated, often tacit and latent, "and sometimes wrong (people aren't sure what they want)" (p. 3). This statement suggests that all design begins with needs as well as the goals described previously. There may be two types of needs — those of the system owner who contracted for its development and those of the operator (when he or she is not the owner). The needs of the owner are to understand and control the world that the system controls; the needs of the operator (the one in immediate contact with the system) are to understand how the system functions in relation to the world it controls and the operator's role in relation to the system. The operator need not be concerned about control; the computer in the system handles that. The owner need not understand how the system controls its world; the owner simply needs to know that the system performs its functions.

Because the owner's needs are simpler than those of the operator (the operator needs to understand the symptomatology of system functioning, integrate diverse items of information, and interpret them), software provisions for the operator are much more complex and difficult to accommodate. If the owner has little understanding of operator needs, the former may be satisfied with a design solution that may fail to satisfy the latter's requirements. Under these circumstances, if operator requirements are not satisfied, the owner may be less satisfied with the system output.

Design usually proceeds under conditions of great uncertainty; often answers to critical questions are not available when needed. Nevertheless, design must proceed by making assumptions that may later turn out to be less than adequate. Unresolved issues may remain unresolved until later in design because answers are not immediately available. The individual components of the design problem are highly interdependent, but each must proceed even if difficulties with one component are not immediately resolved. This may create later difficulties in synthesis.

For all but the most insignificant problems, several different engineering disciplines are required, which makes communication critical to design. Various stakeholders in the design problem view it from their individual perspectives and interests, and these have to be resolved.

Design generation and synthesis, viewed from a great height of abstraction, can be thought of as a control process. The DR study referred to previously is aimed at achieving that control.

Design Rationale is, at any one time, (a) an expression of the relationship between conceptual orientation and design constraints; (b) the topic that pushes the design and a notation to describe that topic; (c) a method of design in which the reasons for the design are made explicit; (d) the documentation of the preceding; and (e) an explanation of the reasons for the design.

Notice the terms that imply control: *expression* (which means making covert processes overt), *logic*, *justifies*, *notation*, and *documentation* (again rendering covert conceptual processes overt), *explicit*, and *explanation*.

The DR research (because the methodology is presently only tentative) is being performed by a number of different organizations and individuals. Fundamental to the work is an attempt to capture what real software design consists of. The following extends the discussion in Chapter 4 to software design. As one part of the research, 10 design meetings from four projects in two organizations were videotaped, transcribed, and then analyzed using a coding scheme based on QOC (*q*uestions, *o*ptions, *c*riteria) methodology. It was found that only 40% of the decision time was spent directly on design issues, another 30% was spent on examination of progress, using such devices as walkthroughs and summaries, and 20% was spent on coordination and clarification of ideas.

Singley and Carroll (1996) focused their psychological analysis of design on the relationship between design features and user psychology. They viewed design as taking place in a rich historical context. This means that the design process can be conceptualized as essentially redesign (e.g., specialization, extension, simplification) of existing design outputs. Redesign is driven by a simple heuristic: Within the constraints of the design task, try to eliminate or mitigate those factors involving negative psychological consequences to users while enhancing positive psychological consequences. This places the user-task scenario at the heart of the design process. The new design requirement will, however, have new features, and this can change the nature of user-tasks to be performed so the psychological analysis that has driven the redesign may no longer be valid. Some critics have said that this design philosophy requires iteration and early feedback from users. However, this also suggests that design is something of a blind search that is informed by repeated evaluations of successive design artifacts. (This may be what rapid prototyping is all about.) The gap between analysis of existing products and new design is essentially a creative (i.e., intuitive) act, but the intuitive aspect can be modified (constrained) by bringing psychological knowledge from various sources to bear. In good design, constraints are constantly being applied.

Singley and Carroll conceptualized five modes of reasoning that drive design through what they called the *task-artifact cycle*. These modes are scenario-based because they analyze the interactions of users with design outputs in task settings. The five modes are:

1. *Hill climbing* from a predecessor artifact. Design decisions may be motivated by an analysis of the strengths and weaknesses of an existing

system in terms of functionality, interface techniques, or tasks implied by these. A perceived problem may be fixed or a new feature added. For example, spreadsheet development has been largely an evolutionary process involving hill climbing.

2. *Process modeling.* An interactive process may be observed that can serve as a model or metaphor for a new situation. The design task is to abstract the principles of the model and implement these in the target situation. An example might be modeling the operations of an ATM on the operations of a human bank teller. The modeling process involves mapping, but also considerable interpretation on the part of the designer.

3. *Assessing an artifact genre.* The preceding two modes involve fairly concrete mapping from predecessor to new design. However, the analysis may take place at a higher level of abstraction, in which predecessor designs are viewed as instances of a genre or class of system. This is an analysis of the strengths and weaknesses of an entire class of design products, which requires a fairly mature technology.

4. *Envisioning scenarios.* The preceding concept modes can define goals and constraints, but they do not suggest concretely what form the new design will take. The designer must now perform creative problem solving by proposing a unique configuration of figures and functions to satisfy design constraints. In scenario environments, one or more user scenarios are developed. The essence of this is simulation, which may be implemented in various ways: mental images, paper and pencil, or computer mockup. The essence of the scenario is what users will do with the proposed design. The scenarios may be developed as textual narratives and then analyzed.

5. *Formative evaluation.* In this process, scenarios are translated into code. As the new design proceeds, it can be submitted to formative evaluation, which involves nonexperimental testing.

There is an implied sequence in the preceding modes of thought. The first three are what one would expect in preliminary design, envisioning scenarios are detailed design, and formative evaluation is included in later detailed design.

The preceding are not the only software design methods theorists have conceptualized. (It is interesting that design theory has not been, in the authors' experience, a favored topic for hardware designs. It may be that the more extensive possibilities in software design, with their emphasis on cognitive functioning, are more conducive to software design theories.)

Potts (1996) categorized the new design methods as *structured, object-oriented,* and *special-purpose* and *hybrid* methods. Each method concentrates a designer's attention on different issues. Despite these differences, they have three things in common:

1. As one would expect, each method has a design philosophy or strategy with which to attack the design problem. Such a philosophy is simply an implicit view of what software design is. Structured analysis conceives of design as recursive, in which a holistic, functional description of the problem is decomposed into smaller subproblems. This is the general design philosophy described in this chapter. Object-oriented design divides a system into a number of objects (subsystems), each of which stands for the operations and data needed to perform those operations.

2. Methods also have heuristics (tactics) in addition to the global strategy suggested by the overall design philosophy. As one would expect, methods vary widely in the tactics they suggest.

3. The various methods differ in the kinds of products they assume will be produced, although all models yield raw code. Before the code is developed, however, there are major differences, representations, models, descriptions, or specifications of the system that eventually lead to code.

Regardless of all the supposed methodological differences, information as categorized in Table 6.1 is requested. These categories were developed on the basis of designers talking about design in a number of protocol studies (Gruber & Russell, 1996). Information sources included people, documents, databases, CAD tools, spreadsheets, and notes. The scope of this information is very broad. No single model of the design process (e.g., design as argumentation, design as decision making, design as constraint satisfaction) dominated the information requested. The disadvantage of using a model as an organizing strategy for a particular problem is that it only ensures that questions appropriate to the model are asked.

Nevertheless, the effect of the information requested in Table 6.1 is to construct a model of the world with which the system will interface. (Note that this is a model of the world; the model may not describe the totality of the world or all its aspects, only those the designer considers pertinent to the system under development. This last is a judgment call, which is why the model may occasionally be incorrect.)

Two essential elements of all design (hardware as well as software) are that one constructs design models and then evaluates them. The designer relies not only on predecessor models, but in the course of satisfying the immediate design problem creates another model, which becomes part of the design history of a class of system. In HCI, the major medium for influencing design evaluation (at least in early and intermediate phases) is the guideline. The major factor influencing construction of the model is the history of successful, existing interfaces. Needless to say, the construction-evaluation process is inherently iterative.

Potts (1996) suggested that designers typically apply the following process. For any proposed design option, they should ask what questions

TABLE 6.1
Categories of Information Requested About Designs

Category	Questions
1. Requirements	What are the given requirements?
	Is this constraint a requirement?
	Can you give more detail about this parameter of the operating environment?
	Can I assume this fact about the operating environment?
	What are the requirement constraints on this parameter?
	Is this parameter constrained by external requirements?
	What is the expected behavior of this artifact in the scenario of use?
	Should I assume that this functionality is required?
	Can I modify this requirement?
2. Structure/Form	What are the components?
	What class of device or mechanism is this part?
	What is the geometry of this part? (qualitative)
	What material is this part made of?
	How do these components interface?
	What are the locations of parts, connections, etc. (for constraint checking)?
	What are the known limitations (strengths) of this part/material class?
	What affects the choice of artifact components?
3. Behavior/Operation: What Does It Do?	What is the behavior of this parameter in the operating conditions?
	What is the behavioral interaction between these subsystems?
	What is the range of motion of this part?
	What is the cause of this behavior?
	What are the expected failure modes in the scenario of use?
4. Functions	What is the function of this part in the design?
	What is the function of a feature of a part in the design?
5. Hypotheticals	What happens if this parameter changes to this new value?
	What is the effect of this hypothetical behavior on this parameter?
	Adapt equation to this changed parameter and recompute. What will have to change in the design if the parameter changes to this new value?
6. Dependencies	What are the known dependencies among the parts?
	What are the constraints on this parameter?
	Is this parameter critical (involved in a dominant constraint)?
	How is this subassembly related to this parameter?
	What is the source of this constraint?

(Continued)

TABLE 6.1
(Continued)

Category	Questions
7. Constraint Checking	Is this constraint satisfied?
	Does this structure have this behavior that violates this constraint?
	What are the known problems with this design?
	Would a part with this functionality satisfy this constraint?
8. Decisions	What are the alternative choices for this design parameter?
	What decisions were made related to this parameter?
	What was an earlier version of the design?
	What decisions were made related to satisfying this constraint?
	Which parameter, requirement, constraint, or component should be decided first?
	What design choices are freed by a change in this input parameter?
	What alternative parts that satisfy this constraint could substitute for this part?
	Where did the idea of this design choice come from?
9. Justifications and Evaluations of Alternatives	Why this design parameter value?
	Why is design parameter at Valve V1 instead of normal Valve V2?
	Why was this alternative chosen over that alternative?
	What is person P's evaluation of these alternatives?
	Why not try this alternative?
10. Justifications and Explanations of Functions	Why is this function provided?
	Why is this function not provided?
	Why can't the current design achieve this new value of this functional requirement parameter?
11. Validation Explanations	How is this requirement satisfied?
	How is this function achieved?
	How is this functional requirement achieved?
	How will this part be maintained?
12. Computations on Existing Model	Compute a parameter value given other parameters.
	What are the trajectories of parameters?
13. Definitions	What does a term in the documentation mean?
14. Other Design Moves	What information is expected to be in documentation (e.g., equation or diagram)?
	Change this requirement constraint and update design. Have all the arguments for/against this alternative been checked?

Note. Derived from an analysis of designers talking about designs in the protocol studies surveyed by Gruber and Russell. Taken from Moran and Carroll (1996).

the option answers. Once those questions are identified, they should ask what other options might be included as answers to the questions. That process involves asking if there is a higher order (broader or more inclusive, or consequent) question. For any list of options, the designer should ask what the criteria are for making a choice among options. To summarize, design becomes a matter of exploring the decision space with questions, options, criteria, and consequent questions. The Q-O-C-C process, as it is called, is illustrated in Fig. 6.2.

Suppose someone suggests a pull-down menu for a particular application. If this device is an answer to some question, what is the question?

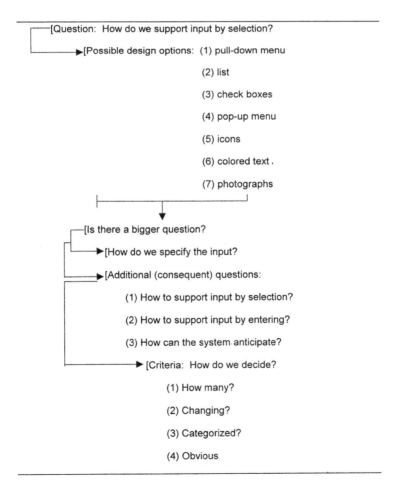

FIG. 6.2.   The Q–O–C–C process (from Moran & Carroll, 1996).

Prototyping and Iterative Design

Prototyping is a way to test design. As a test, it is like any other test: the presentation of stimuli to subjects and measurement of their performance and preferences.

We have already discussed prototyping and its related test activity, usability testing (UT), but a few additional words should be said.

In the context of software design, prototyping has one significant value: It brings the user more directly into the design process. In the past, skilled hardware designers thought about or considered the user and the effect of the design on user performance, but they did not ordinarily try to contact users for specific opinions on design (except as SMEs).

There are three aspects to prototyping: (a) developing the prototype, which should be built easily and cheaply and should be easy to modify to implement redesign; (b) development of the test situation, which means selecting the design features to be presented, selection of the test population, measurement in the test situation, and analysis of whatever data are collected; and (c) determining which design features should be modified as a result of the prototype test. The second aspect is testing, and the third aspect is design evaluation and deduction based on test results.

Liu (1997) called out prototyping as an individual design stage, but this is a misconception; prototyping can be applied in every design stage, including preliminary design.

Prototyping can be performed in various ways: rapidly (hence the term *rapid prototyping* [RP]) or in a more leisurely manner; formally or informally; with or without test controls, the controls involving such aspects as deliberate selection of subjects and prior testing of the measurement instrument. Rapid informal prototyping can be something as simple as buttonholing a colleague in the hall and asking, "Joe, would you take this test for me?" or "Would you look at this design and tell me what you think?" Do this with a number of colleagues and the entire test can be concluded in 1 day. The question that has been explored in the literature (Virzi, 1989; Wilson & Rosenberg, 1988) is whether such a slapdash test can produce the kind of information desired.

A prototype test is, in the last analysis, a stimulus to the designer that directs his or her attention to redesign of some feature not previously thought of or about which there was some doubt. Or it can be considered a sort of informal verification test: Is this feature I have included really worthwhile? What the prototype test consists of depends largely on what the designer hopes/expects to get out of it. All of this permits a great deal of variation. The prototype test may even assume the character of a fishing expedition, in which anything that results from the test may (or may not) be useful.

It should be remembered that there is no requirement that the designer accept the results of the prototype test. In any event, the designer must interpret the test results to find something meaningful.

Obviously a major value of RP is that it is rapid; design is always under the Damocles sword of insufficient time. The theorist (and here we adopt that role momentarily) would prefer that RP be conducted more formally (which means more slowly) because we assume that the more formal the test, the more information it will provide. However, this is true only if we believe the test is to provide information; if it serves primarily as a stimulus to the designer, the amount of information it provides or does not provide is less important.

Ideally, what we want the prototype test to tell us is: (a) Can the operator/user employ the equipment (SEP) or selected features of the SEP without undue strain? and (b) What features are most preferred or disliked, and why?

The measures that can be applied are those of any other test in which performance is recorded: observation, performance measurement, interviews, audio/video recording, scales, and questionnaires. Liu (1997) indicated that a number of software tools are available to support prototyping: Motif for the XWindows environment, HyperCard and MacApp for the Macintosh, Toolbox and Visual Basic for the IBM-PC, and widget sets for DECWindows and Microsoft Windows, and Powerbuilder, C++, and other Hypermedia graphical user interface tools. No doubt, by the time this is read, there will be new ones on the market.

Prototyping (RP) and usability testing (UT) are linked methodologically. To the extent the RP has clearly specific test goals, test plans, and test procedures, it may be indistinguishable from UT. Subject performance data in the form of response time and errors, successful completion of the prototype task, quantified (by scaling) measures of opinion (preferences), debriefing following task performance, audio/video recording — in effect, the formal prototype test or UT — is not much different from an experiment and may be an experiment if alternative forms of the prototype are presented and subjects are asked to perform with each. The interesting thing about RP and UT is what the designer does with the test results. This is not as mysterious as the original creative design act because the test data provide clues for redesign, but it still requires some creative translation of the test data into design modifications.

For commercial systems, which are addressed to multiple consumers (e.g., mobile telephones), emphasis in the RP and UT is placed on preference data, which are linked to aesthetic features of the design; when the SEP has a single owner (commercial or governmental) and is, moreover, complex in its operation, greater emphasis is placed on determining whether subjects can use the device.

For testing of general commercial use products, the greatest variety of consumers should be selected as test subjects. Selection criteria are particularly important where subsets of the general population (e.g., children, adolescents, bald men, the elderly) are targeted by the equipment. For single-owner SEP, which require specially skilled personnel as operators, only personnel possessing those skills (e.g., pilots) can be utilized as test subjects.

The prototype constructed for presentation to subjects may not be complete in the sense that it contains all the features that the operational device will have. Therefore, prototype testing and UT cannot replace a formal operational test (OST) in the operational environment or in a simulation of that environment. Considering its budgetary constraints, management may wish to substitute RP or UT for OST; this is not acceptable.

Part of the formal prototype test or UT is the matter of documentation for later analysis. Designers and programmers are notorious for incomplete documentation of the processes they use to arrive at design decisions. As a result, "what you see is what you get." The documentation of any evaluations and tests performed during development cannot help but produce a more effective device if only because documentation usually requires closer analysis of the problem by the designer.

Interface Development

Again, it must be emphasized that interface development must begin in parallel with other system development processes. That is because the interface must mirror the internal functioning of the system and the world that the system controls or examines. Because the computer now controls the system, the operator's role is now as a monitor to ensure that the system performs to preset limits, diagnose impending or actual malfunctions, and analyze and correlate/integrate information from diverse sources. To design the interface properly, its designer (engineer or HF design specialist) must know almost as much about the internal workings of the system and world dynamics as do other engineers.

The interface may be developed late in the system development sequence, but the information on which that development is based must be gathered much earlier.

As we saw from Table 6.1, design proceeds from questions asked by the design team and the answers to these questions. That is as true for interface design as it is for functional design. The questions we would ask are listed next; they are somewhat general because we are not asking them of a specific system.

1. What are the major internal components of the system and how do they function (list the operations they perform sequentially as in a flow diagram)? How are these components interrelated?

2. What does the designer want the operator to do in relation to these components and the overall functioning of the system? Here the designer must refer back to the task analysis (TA and CTA) developed by analysis in the initial Requirements and Task Analysis phase. There are a number of possibilities for the operator: monitor performance only; if any malfunction appears imminent, take emergency action; select one of a number of operating options, if these exist; shut the system down when it appears necessary; diagnose the system problem and perform corrective maintenance; call for more specialized maintenance personnel. The functions and tasks assigned the operator will greatly influence the information to be presented on the interface; the more operator functions to be performed, the more information may be required. Again, it must be noted that the designer always has a choice about what and how much information should be presented to the operator. This decision may be the most important one that can be made during interface design.

3. What are the quantitative values the operator should look for while monitoring the system operation? These are the parameters within which the operator will work.

4. What are the various ways in which each major component alone (or in combination with others) can fail and what will the failure symptomatology present in the event of failure?

(In the event that the reader thinks that the development of this information will be inordinately difficult, it should be pointed out that, traditionally, reliability engineers perform what they call a "failure modes and effects analysis" to supply just this type of information for their reliability predictions. It is also very possible that a reliability engineer will be part of the design team for the system.) Moreover, many of the questions listed in Table 6.1 deal with just these items.

5. What information should be displayed on the interface and how should that information be displayed (i.e., types of displays and controls, how they are laid out, etc.)?

The first set of questions about the functioning of the internal components is designed to establish the content of the interface information to be provided. The engineer and reliability specialist will have primary responsibility for answering these questions, although the interface designer should be cognizant of what is discovered.

Once the content of the information has been established, a second set of questions deals with how the information should be presented. The HF design specialist should play a major role in answering these questions because whatever is decided on as to information presentation may have to be modified by purely behavioral considerations, such as how much information personnel can assimilate in a time period. The answers to this second set of behavioral questions should be derived from the HF research literature and from previous and concurrent prototype testing.

The behavioral questions are:

1. Does the amount of information to be presented exceed the perceptual/cognitive capabilities of the operator? We have in mind that the interface can easily become a busy place for the operator. Two conditions must be considered: normal routine monitoring and emergency situations. It may be necessary to drop certain items of information, in which case a judgment must be made as to the relative importance of individual information items.

2. What information items must be integrated or related to each other by the operator, or can the software perform the integration and present the results? Can integrated displays, which summarize several parameter values, and predictor displays, which extrapolate trends, be utilized? If information integration, which includes interpretation, of course, must be performed by the operator, how difficult will this process be?

3. Are there any perceptual/cognitive biases noted in the HF literature that must be considered in the interface design? These biases may involve such design aspects as the way in which displays (and controls) are laid out, how information is highlighted or enhanced, and the sequencing of displays to simulate or mimic internal system operations.

We assume that during interface development there are considerable prototype or usability testing to determine how well operators are able to function with the interface. What is presented to operators during RP or UT is not only tentative interfaces, but also representative situations in which the interfaces are used (i.e., monitoring problems, symptom diagnostic problems, problems of integrating/interpreting information). The term *test* has a special meaning here; the test is measurement, but measurement of operator capability to solve problems occurring in routine and emergency use situations. For example, subjects may be presented with examples of symptomatology and asked to say whether they represent correct system functioning, an impending failure (and of what type), and to diagnose an actual malfunction. The stimuli can be presented either on paper (a static test situation) or electronically on the interface terminal (dynamically), in which latter case the operator subjects can observe the buildup of a problem and determine what to do about it in real time. Operator responses may be written or expressed in terms of physical actions (by pressing a button, entering an input command via keyboard, or actually manipulating a simulated control).

If subjects cannot make the proper diagnoses, if they make consistent errors, or if they rate performance with the interface very difficult (and for specific reasons), designers have to make necessary modifications. The criterion of interface adequacy is not only an engineering one, but in addition (and equally important) one of satisfactory operator response. As indicated elsewhere, performance criteria must be established before the test(s) to indicate whether the interface is adequate. Assuming, for example, 20 simulated problems are presented, is a correct score of 15 problems solved adequate to demonstrate interface functionality? Functionality for the interface is whether operators can use it properly because the interface exists only to permit the operator to perform designated functions. Behavioral functionality must co-exist with physical functionality.

This is something that engineering and management may not wish to accept because it means that merely installing controls and displays and ensuring that they work physically is insufficient to define interface functionality. The HF design specialist may have to work hard to convince engineers of this point.

In this discussion, we have concentrated on the purely behavioral aspects of interface design. There is a vast literature on types of displays, input devices, and so on, for which the reader can do no better than to consult the admirable summaries in Salvendy (1997).

Final (Verification) System Testing

At this stage, the computer software system is considered by its designers to be completed. Initial production has begun, and management believes that the system can be transferred to its owner or sold commercially.

The designer's task is not complete, however, until the system has been exposed to verification testing, either under engineering control or by placing the system in operational test use with a number of potential using companies and/or personnel. This end-of-development process has been described previously, and there is no need to repeat it except to emphasize that verification testing is a requirement that should not be avoided. Final system testing should be a formal process, with particular emphasis on placing the system into the operational environment or some reasonable simulation of that environment. The complete system must be tested, not bits and pieces, as may have been the situation previously.

Formal evaluation criteria should be applied to the final system test, with particular emphasis on criteria for evaluating operator performance. Unless the system is very simple, few criteria are dichotomous (and, consequently, simple) — the system either succeeds or fails, as in the case of a missile interception system, which either shoots its adversary down or does not. Most systems function with some degree of success or failure, which must be clearly specified. This is particularly true in the case of the interface whose adequacy is measured in the context of operator/user variability.

OST or any comparable test may reveal certain inadequacies that will have to be corrected by the developer. It is probable that every system is placed into operational usage with minor defects.

COGNITIVE AND MENTAL MODELS

The focus on the operator/user in software design has led to the development of what are termed *cognitive models*. These describe in quantitative form the way in which personnel make use of the interface to perform tasks.

At the same time and for the same reason, there has been much interest in mental models, the concept held by the designer and by the operator/ user of how the system functions.

Cognitive Models

Our discussion is adapted from a review by Liu (1997), who saw these models as a way of making quantitative HF inputs to design at an early stage.

Cognitive models are software engineering methods that may be of use in specifying interface requirements, such as menu lengths. In the following discussion, the reader should keep in mind two questions: (a) Are these models valid descriptions of the operator's conceptual and motor actions; and (b) If they are valid, are they useful to software design and are they in general use in interface design?

Table 6.2 is a list of cognitive models taken from Liu (1997). There are two classes of models: those based on formal grammars and those based on transition diagrams. Formal grammar models contain rules for producing correct language statements. Examples are Reisner's (1981) grammar for describing human–computer interpretation languages and Payne and Green's (1986) Task Action Grammar for describing interface structure and evaluating interface consistency. Transition diagram models represent system status and operator actions in the form of graphs containing nodes and arcs. Each node represents a system state in which the interface is awaiting some sort of operator action, and each arc, which begins at one node and ends at another, describes an action that the operator is permit-

TABLE 6.2
Major Computational Models of Cognitive Performance

Specification Language or Grammar-Based Models	Models Based on Operations Research Methods
Model Human Processor (MHP) and GOMS (Card, Moran, & Newell, 1983)	Queuing Theoretic Models (Moray, 1986; Rouse, 1980)
Natural GOMS Language (NGOMSL; Kieras, 1988)	Control Theoretic Models (Baron & Levison, 1977; Pattipatti & Kleinman, 1992)
Cognitive Complexity Theory (CCT; Kieras & Polson, 1985)	Systems Analysis of Integrated Networks of Tasks Models (SAINT; Laughery, 1989; Siegal & Wolf, 1969)
Critical Path Method-GOMS (CPM-GOMS; John, 1988)	Critical Path Network Models (John, 1988; Schweikert, 1978)
Command Language Grammar (Moran, 1981)	Queuing Network Models (Liu, 1994, 1996, 1997)
Cognitive Grammar (Reisner, 1981)	
Task-Action Grammar (Payne & Green, 1986)	

Note. From Liu (1997).

ted to take when the system gives the user a green light. An example of such a model is Wasserman (1985).

Most information-processing models have not been developed for use in interface design. An exception is the Model Human Process (MHP), which was developed by Card, Moran, and Newell (1983). The purpose of this model is to make quantitative predictions about operator performance times.

Much better known is the GOMS model (goals, operators, methods, and selection rules). This model, which was also developed by Card et al. (1983), analyzes cognitive tasks in terms of methods for achieving task goals; it includes rules for choosing among various methods of accomplishing those goals. The performance measure it employs is an estimation of the time it takes to complete these tasks. GOMS has spawned a number of variations. One, which is called natural GOMS (Kieras, 1988), permits the user to describe HCI interactions in a language similar to that of computer programming languages. Another variation is called cognitive complexity theory (CCT; Kieras & Polson, 1985). Then there is the critical path method (GOMS; John, 1988), which extends GOMS to the analysis of parallel activities. Command Language Grammar (Moran, 1981) is a grammar based on a TA method.

Other models are based on operations research and industrial engineering methods. Liu (1997) postulated four classes of these: queuing theoretic models, control theory models, discrete network models, and queuing network models. Queuing and control theory models assume that the human information-processing system (whatever this is, biologically or functionally) has a single, central processing unit that permits the cognitive function to switch its capability among concurrent competing task demands according to some scheduling or resource-allocation strategy.

There is also the most famous of the task network modeling methods (SAINT, systems analysis of integrated networks of tasks; Laughery, 1989). This method models human–computer interaction as a sequence of tasks. Alternative sequences to accomplish a goal form a network; parallel sequences represent alternatives. Schweikert's (1978) critical path network also represented information processing as a task network; its parallel pathways represent concurrent rather than alternative processes.

There have also been attempts to combine models. John (1988) integrated the critical path method (CPM) with GOMS to form CPM–GOMS. In both SAINT and CPM, only one process in the network can be active at one time; hence, both are discrete network models.

Queuing network models represent cognition as a network of distinct servers, each of which is responsible for a single function or task and has a waiting space for tasks that cannot be immediately processed. Individual task aspects are serviced by individual servers, both in parallel and concurrently. When task demands are high, there may be multiple queues of task demands awaiting service. This type of model allows processes to be active concurrently; hence, the model can accommodate alternative, concurrent, parallel, and sequential processes (Liu, 1994, 1996, 1997).

All of these models make someone who is less than an initiate somewhat uneasy. Admittedly, each follows a somewhat different strategy, but it seems unlikely that all present equally valid or useful pictures of conceptual functioning. In any event, the profusion of models requires comparison among them and some sort of validation process. Eberts (1997) reported that the GOMS predictions have been tested against empirical data and predict accurately at an 80% to 90% level; but validation has not been reported for the other models. The lack of validation is like that for other human performance models that are not organized specifically for cognition.

The models are very molecular (see Eberts, 1997, for examples of GOMS use) and are consequently quite complex. Eberts made the point that the particular values the models provide depend to a certain extent on the skill of the model user. If, to apply a model, one must be an expert in modeling, this requirement will reduce the actual use of the models significantly.

One can think of these models as having utility (whatever their validity may be). Task network models have proved useful in research (e.g., Siegal & Wolf, 1969), but how much use is made of them in actual design is unclear. There are repeated references in the literature to SAINT and micro-SAINT (a refinement of the original) and this is hopeful.

One can think of these models as sophisticated efforts to bridge the human–technology gap, but the question remains, how well do they do this? They can (or at least GOMS can) predict text-editing performance successfully when equipment details and operator tasks are input to the model. Presumably one could vary these equipment details and operator tasks and compare the variations to determine the most effective. This would be a form of prototype testing using only the model (no human subjects). If this were possible, the models would be a useful application to design. However, the equipment and task details must be developed first by design personnel and then input to the model before any answers can be secured from the model. The human performance model is a great potential asset to design, but, as usual, humans (design personnel) must do all the hard work before the model can be energized.

Currently, there are attempts by the U.S. Navy to design ships with reduced manning, and human performance models are part of those efforts. However, as yet there is no information about whether such models have been useful in this context.

There is, then, a great need to survey the experiences of design personnel to determine what is done (if anything) with models in actual design. We do not know which models are used in actual design or how frequently, or how effective they are in aiding design. On a more general note, for a discipline that is a quasi-engineering one (and should therefore be rather hard headed), little information is collected about the real-world experience with the HF methods that have been developed to aid design. Perhaps one can blame this at least partially on the academic traditions to which HF (through its predecessor, psychology) is so strongly attached. It appears as if little in HF has value unless it has been derived from classic laboratory experimentation. The empirical side of HF (the side involved in design, system development, operational use) suffers from the reluctance to look forthrightly at what we do. There is a lack of feedback from the practitioner side of HF, which significantly reduces its effectiveness; without feedback, there can be few attempts at improving this aspect of HF. The concept of feedback is enshrined in HF theory and methodology, but apparently only as it applies to HF research.

Mental Models

The mental model is a parallel to the cognitive model, but differs from it significantly:

1. It resides within the human and is purely mental. Therefore, it cannot be easily manipulated, although it is possible to study it empirically;

2. It is entirely qualitative;

3. As a representation of the system, it is often incomplete, vague, and confused; and

4. The mental model almost certainly has a significant effect on the performance of the system because it is a primary factor determining how well the operator/user performs.

The operator/user of a computer system develops a mental model of how that system functions based largely on a manual describing how the computer works, with instructions for the user to follow. Most company-produced manuals are completely inadequate. However, they are supplemented by manuals written by personnel who understand the system and can verbalize that understanding.

The designer also has a mental model of how the system functions; he or she cannot develop the system without creating one. The designer uses that mental model as an input to the technical writer, who develops the written manual given to the user.

It is our impression that it is rare for the designer him or herself to write the user manual because most engineers seem to lack the verbal fluency needed. However, because there is a communication process involved in providing inputs to the manual writer, the possibility of confusion and misunderstanding always exists, although the designer presumably reviews the manual before it is published. It must also be remembered that the mental model is not verbal; it is graphic, and thus must be translated into words. The transmission process is therefore fraught with peril for the user. As a result, a cottage industry has developed, in which independent experts write books (like *DOS for Dummies*) explaining the system to users. The reality is that many users have an inadequate model of their systems, which causes them to make errors, experience stress, and swear at the software developer.

It is entirely too easy to blame the unfriendliness of the computer on inadequacies in the user's mental model. The problem is one not only of the transmission of concepts from the engineer, but also of individual differences in user intelligence, skill, age, and personality, which exacerbate the problem.

Norman (1983) went so far as to say that, if the user's model does not agree with that of the designer, the user will inevitably experience difficulties. He suggested that the designer should anticipate the user's model and incorporate features into the system that will modify the user's model to become more like that of the designer. How this can be done when the user requires a completed system before he or she can develop a mental model of that system is unclear. Theoretically, therefore, the software designer should have two mental models: one of the system as he or she understands it and a second model of the operator's mental model. This may be asking too much of the designer and specialist. However, the notion that the designer should anticipate the general classes of problems users will encounter with the new systems and the kinds of questions they will ask is an excellent one.

It is necessary to differentiate the specialized professional operator from the general consumer/user of PCs. The former is more highly experienced and skilled and generally receives special training before operating the system. The latter is generally left to the tender mercies of instruction manuals, which are usually grossly inadequate. One would expect, then, that professional operators would have a more complete and accurate mental model than general users.

PC users do have commercial (for-profit) training courses available to them for either a word processing or spreadsheet program. The cost of this low-end training is approximately $60.

Because the user's mental model is in the brain, it cannot be manipulated directly. However, it can be studied empirically by training subjects in system dynamics and then asking them to reproduce their mental model of the system by talking or writing about it, providing demonstrations, answering questions, and so on. It is even possible to introduce experimental variables (e.g., special system features) to determine the effect of these on the model.

Many software engineers assume that everyone thinks as the engineer thinks, disregarding that most users are not engineers. Moreover, the engineer has the advantage of having developed the system so that he or she has experience on his or her side. Engineers also think as a consequence in a sort of conceptual shorthand. Describing that system in the same shorthand manner may lead to user difficulties. It helps if the engineer can imagine him or herself as a user without any engineering knowledge.

Inevitably, theorists will attack the problem, and a number have done so (Bainbridge, 1981, 1991; Goodstein, Anderson, & Olsen, 1988; Moray, 1986; Norman, 1983, 1986; Rasmussen, 1986; Rasmussen et al., 1994). There are apparently two ways to develop a mental model: reading a description or observing a demonstration of a system, and hands-on experience in operating the system. It is also possible that the operator relates the new system to other, similar systems he or she has worked with. Nonetheless, it is almost inevitable that most mental models are vague, unstable, unscientific, incomplete, and, in general, inadequate.

Moray (1997) pointed out that mental models for emergency situations are the most difficult to develop because everything the operator has learned about routine operation is now radically changed. Emergencies, such as the one at Chernobyl, are infrequent affairs; unless one has many years on a particular system operation, there may be no progressive buildup of experience to assist in the interpretation of the emergency symptomatology. An analysis such as the reliability engineer's failure modes and effects analysis, which is a systematic examination of at least the most likely failures and their associated symptomatology, could be very helpful. However, it is not clear how often such an analysis is performed, much less disseminated to designers and users.

If the operating crew of a system like a process control facility have differing mental models of the system, the same information presented as symptomatology may be interpreted by each crew member differently, which could be dangerous. The idiosyncratic character of the mental model adds to the difficulty of producing a common conceptual structure; if there are five operators, there may be five different models.

Theorists lean toward two different approaches to the understanding of mental models (Moray, 1997). The first emphasizes factual knowledge as expressed in displayed information. A second approach emphasizes not so much knowledge as the choice of what should be done next. These two approaches are not dichotomous; it is probable that all mental models utilize both strategies to a greater or lesser extent. The second approach emphasizes active exploration of the system (hands-on), and the installation of automatic devices that eliminate the need for hands-on control of the system may damage the second strategy.

DESIGN GUIDELINES

Previous chapters have emphasized the importance of guidelines as a means to control design and enable the designer to check the adequacy of the design. Chapter 5 analyzed the characteristics of the HF literature to supply the materials from which guidelines could be developed. Therefore, our discussion of guidelines in software design is brief.

Guidelines have the same importance in software as in hardware design. A large number of guidelines (e.g., several hundred in Smith & Mosier, 1986) have been developed, some of which have been summarized in general texts like Helander (1988). Table 6.3 provides samples of the guidelines available to software designers. All are useful, but most are quite general. For example, Tullis (1988) suggested that information should be presented in proper sequence — a statement that should be obvious to any intelligent designer. Smith and Mosier (1986) suggested that the designer should use short, simple sentences.

Perhaps if the HF literature reviewed in Chapter 5 is representative of the totality of design-relevant literature, this is the best one can do. There are, of course, highly specific guidelines. For example, one of the Smith and Mosier guidelines suggests that when size coding is used, a larger

TABLE 6.3
Samples of Interface Design Guidelines

Guideline	Goals
General Design (Shneiderman, 1992)	• Strive for consistency • Provide information feedback • Organize action sequences into groups • Provide simple error-handling mechanisms • Permit easy reversal of actions • Enable users to be in control of the system • Reduce short-term memory load
Data Display (Smith & Mosier, 1986)	• Left justify columns of alphabetic data for rapid scanning • Label each page • Maintain consistent format in displays • Display data in directly usable forms • Use short, simple sentences • Use affirmative rather than negative statements • Provide a title for every display • When blink coding is used, the blink rate should be 2 to 5 Hz
Screen Design (Tullis, 1988)	• Use abbreviations appropriately • Avoid unnecessary details • Use concise wording • Use familiar data formats • Use tabular formats with column headings • Arrange related items as groups • Use highlighting to attract user attention • Present information in proper sequence
Color Usage (Murch, 1987)	• Avoid pure blue for text, thin lines, and small shapes • Avoid red and green in the periphery of large-scale displays • Note that all colors are equally discernible • Do not overuse color • Use similar colors to convey similar meanings • Use a common background color for group-related elements • Use brightness and saturation to attract attention • For color-deficient viewers, avoid single-color distinctions
Error Message (Shneiderman, 1992)	• Be as precise as possible • Be positive and avoid condemnation • Be constructive: Tell user what needs to be done • Be consistent in grammar, terminology, and abbreviations • Use user-centered phrasing • Use consistent display format • Test the adequacy of error messages • Reduce or eliminate the need for error messages

Note. Adapted from Liu (1997).

symbol should be at least 1.5 times the height of the next smaller symbol. The guidelines in Williges et al. (1987) appear to communicate more specific design direction. Of course, design guidelines represent the literature distilled by the expertise of the designer. One supposes that the computer-relevant literature from which these guidelines were refined should have been more specific than the HF literature that covers much more than computers. If that were the case, it is even less understandable that software guidelines should be as general as they appear to be. This puts into question the utility of any general literature to provide adequate design guidance.

The importance of guidelines is reduced by the use of prototype testing. User preferences and performance in usability tests serve as specific corrections for the design and mitigate the excessively general nature of verbal guidelines.

DISPLAY DESIGN ISSUES

Most of the previous discussion dealt with software design. Software for operators/users can be utilized only if it is presented (displayed) to the user. Hence, any discussion of software also demands discussion of computer graphics and, to a lesser extent, auditory and speaker/headphone outputs. This is a vast area that can only be dealt with adequately in books devoted to the topic. We recommend the chapter by Bennett, Nagy, and Flach (1997) as a summary of what is known about interface displays.

The major display issues, as Liu (1997) listed them, include screen design (e.g., color and layout), typography, data display formats, types of displays (e.g., object and three-dimensional displays), information visualization, multiwindows, display modalities, error messages and helps, and documentation. Liu provided references for each of these.

A New Type of System

In this section, we consider certain inadequacies of present computer systems and the characteristics of more advanced ones. The design of the computer system is performed in accordance with certain principles that, although universally employed, are not necessarily cast in concrete.

We suggest that:

1. There is a more intimate (direct) relationship between the computer system and the operator because cognitive functions are more involved in

software than in hardware operation. The computer speaks more directly in terms of the information the software provides than does the hardware.

2. There is much less proceduralization in software. Consequently, there is more necessity for integrating (perceptually/cognitively) diverse stimuli.

3. This integration leads to a transmission of information about (a) the status of internal machine functioning, and (b) the status of the external world if the device has been designed to interrogate the status of that world (e.g., sonar, radar, or the stock market).

4. The role of the operator changes in dealing with a computerized system: The operator becomes a monitor of system processes and a diagnostician if anything goes wrong. The operator is also allowed to search for information, collect and manipulate it, store and retrieve it, and transmit it.

This is not to say that the recognition of stimuli and their meaning did not exist in equipment prior to computerization, but that the computer permits the user (and is itself permitted) greater complexity in information gathering, integration, and interpretation.

Technological advances have mounted exponentially since the computer arrived. This has exacerbated the central problem of technology, which is complexity. This has led to difficulties in utilizing large-scale systems and small-scale (PC) computers. Building on Flower (1996), Harris and Henderson (1999), from which much of the following has been taken, cited previous difficulties with the Y2K problem, failure to upgrade the computers of the US Air Traffic Control System, the Internal Revenue Service, and the Bank of America.

At the individual consumer level (PCs were found in 40% of American households as of a 1998 government report), there are comparable problems. Although the PC is very reliable, PC software may malfunction and then users do not know how to fix it. There is difficulty upgrading software.

This question may be asked: Why have the huge technical improvements in computer systems proved so difficult to translate into comparable improvements in computer usability? The problem may be rooted in the following assumptions that determine software design: (a) define a clear system requirement; (b) define a clean architecture that can satisfy these requirements; (c) define clear choices for users at each point where the users interact with the system; and (d) maintain consistency throughout design for ease of maintenance and learning.

Underlying these principles are still more fundamental assumptions:

1. System components must interact in terms of a preestablished harmony or arrangement specified during design;

2. The job of the designer is to discover, clarify, and, if necessary, invent the rules that define that arrangement and then embed them in the computer software; and

3. Users must interact with the system in terms of the language these rules create.

One can hardly quarrel with these principles. They are logical, and the computer is built on logic. It would be hard to imagine how system design could be developed otherwise.

The computer system is conceptualized as a model of the external world that it controls, directs, influences, and describes. The real external world is untidy, meaning that variations of all sorts occur. The computer system can handle these variations, which Harris and Henderson called *particularities*, only by taxonomizing them logically and forcing them into what these authors called *regularities* or terms with logical operations. The external world cannot be controlled until the particularities are first controlled. The external world (which includes the user/operator) must play by system rules or chaos and catastrophe result.

Because the external world contains both regular processes and variations from these processes, the success of the computer system depends on how well the system describes and incorporates both regularities and particularities.

Two points need to be made. Because the computer system models the external world, the designer must learn as much as he or she can about that world before beginning design. Either that or include system provisions that enable the system to retrofit and upgrade once it is determined that the system's world model is less than completely accurate.

The other point to be emphasized is that the operator/user is part of the external world and contributes to its variability and untidiness. Operators/users vary in terms of intelligence, skill, knowledge, and personality. As part of the system designer's learning about the world he or she is modeling, the latter should learn a great deal about the operator/user. This is where the HF design specialist has a particular capability.

One must assume that the designer's modeling process, although prolonged, will never be completely accurate in its description of the real world. Thus, there will always be discrepancies, in which particularities do not readily translate to more global regularities.

A translation process is required. Computers can work only in terms of the regularities they have been built to handle. They can respond only in the ways external situations fit these preordained regularities. If the designer has conceptualized rules that do not exactly fit the external world situation, the system will not function optimally.

Computers require users to map particularities into regularities, after which the computer can proceed. The computer system does not deal with the particularity, but only with the regulatory rule into which the latter has been forced. For example, If the system can only deal with animals, vegetables, or minerals, this means that when the name of a flower surfaces it cannot be dealt with as a flower, but first must be taxonomized as some sort of vegetable.

This has unfortunate consequences:

1. Some particularities cannot be accommodated by systems or discussed within them;

2. New regularities and difficulties with old ones cannot even be noticed by the system because these arise only in the process of mapping particularities into regularities;

3. Changes are difficult, slow, and expensive because the system will not notice or accommodate to problems, so all the implications of new regularities must be anticipated by the system designer;

4. Harris and Henderson placed the role of the system in the context of a higher order organization that has a mission to which the system contributes. Computer systems do not share the mission of the organization; all they have is their explicit regularities. The burden of adapting as necessary to carry out a mission falls entirely on system operators/users.

The system is dedicated to the organization's mission, but does not deal with that mission directly, only through the rules developed to implement the mission. A critical further assumption is that, over time, the mission may change or the manner in which the mission is performed may change. This is the consequence of the view that sees the external world as essentially nondeterministic (i.e., probabilistic and therefore highly variable). When the mission changes or varies to adapt to other external forces, the computer has difficulty adjusting. Indeed, it cannot adjust, so users must adjust.

From the standpoint of the HF design specialist, we can translate this as follows:

1. The user is a particularity of the external world to which the computer is supposed to respond. Variations in user intelligence, training, and skill produce particularities of their own. Even if the remainder of the external world does not change (a dubious assumption), it is almost inevitable that the user will, and the system must accommodate him or her.

2. The most common user particularity occurs when the user fails to understand the requirement to implement a computer regularity. Conse-

quently, the regularity cannot function and an error results; the system halts or performs in an undesirable fashion.

There have been attempts to make systems smarter so that they can notice particularities. For example, they should be able to recognize when the user makes a mistake and should compensate for the error. However, such efforts are often ineffective.

Perhaps the world that the computer designer seeks to regularize is constantly changing in so many ways (large and small) that the computer cannot keep up with these changes.

The external world is one filled with uncertainty, which equals indeterminism, which equals change. Indeed, if the system models the external world, the system must include the same types of uncertainty one finds in the world. How can this be done if the computer system can deal only with regularities?

Evolving systems must maintain coherence and agility. Coherence is achieved by the rules describing regularities, which is necessary if the system is to control the changes in the external world. Agility, which is necessary if the system is to take account of (adapt to) those changes, can be addressed by allowing the user to introduce nondetermined particularities.

Computer systems execute programs that specify processes in a completely deterministic manner. The system programs must mesh with each other according to the preestablished harmony (arrangement) because the individual programs cannot adapt to each other. This produces a highly coherent system, but one lacking agility.

In what Harris and Henderson called *pliant* (i.e., changeable) systems, programs will be augmented or replaced with process descriptions using patterns as described by Alexander (1979). These patterns will specify processes only partially and will be open ended. Each process is the result of many overlapping and interacting patterns. This interaction is not the result of preestablished arrangements, but is developed in the process of using them. Systems are assumed to be adaptive, which means that they learn and change on their own. The system may require or request help if patterns do not mesh well, and users may intervene to push patterns into specific relationships.

As opposed to the traditional deterministic system, in pliant systems, patterns will be performed (enacted) and interpreted by finding ways of relating the patterns to the specifics of the individual case. The performance of the process determines which aspects of the situation are relevant and how they should be organized to satisfy the needs of the patterns being performed. The system may seek help from users, and users may intervene to influence how the patterns are performed. To permit meaningful user interventions to system processes requires designers to introduce what we call *entry ports* or points at which users can communicate directly with the software to modify a process to accommodate an important particularity. The user intervention will introduce a degree of uncertainty into system regularities. However, once the process proceeds, the uncertainty is incorporated into the software and its use regularized.

All of this is very abstract, and one wonders how the concepts can be implemented in view of the strong influence determinism has over design. Nevertheless, it is refreshing to think about computer systems that are not bound hand and foot to deterministic principles. Pliant systems, if they can be achieved, will be much more sophisticated than our present systems. However, if this is to happen, it will require much more of our software developers.

The preceding discussion has caused us to emphasize certain things that can be implemented even in present systems. The notion of the system as modeling the external world and the user means that much more attention must be paid to learning about that world and the user before system development begins. The learning process has not been as extensive as it should be because of constraints in time and money, and because in many cases developers do not really believe the learning process is important to design. Because the user is a contributor to the untidiness of the external world, the HF design specialist must be able to provide much more information about how the user's characteristics influence system operations. This will require much more research about user characteristics, not only in response to an already established technology, but also relative to how these characteristics can influence technology.

7

The User

It should be obvious that technology does not exist as an abstraction; there are always systems, equipments, and products (SEP). In addition, and this is the topic of this chapter, there is always a human who uses that technology. Technology that cannot be used is an oxymoron, a contradiction in terms. The relationship between the human and technology is reciprocal: The human has always been influenced by technology, and the human influences that technology by imposing certain requirements and constraints on how the technology is created. The physical form of the human dictates some aspects of the technology (e.g., the shape of the arm and the hand had to be taken into account when designing the rifle), and the human's cognitive (e.g., creative) capabilities determine what new computerized technology will be developed. We have postulated that the human–technology relationship is the essence of HF; to understand that relationship, it is necessary to learn not only about machines, but also about the human or, as he or she is termed in the literature, the *user*.

Everyone is a user in the sense of operating devices or being influenced by them. However, the user is not a monolith. Whom we call the user is determined in large part by our perspective of him or her. There is the user as an abstraction, as described in texts such as this one. There is the user as an object of measurement in design, as subjects in prototype (RP) and usability (UT) testing. There is the user (as an ideal) whom we say we must consider and whose needs and limitations we attempt to incorporate in design. There is the user as the subject of study (to answer the question, who is the user?). There is the user who is the agency or company that desires, funds, and uses the development of a new system. As suggested in Chapter 5, the specialist and designer who make use of HF research outputs are also users for that aspect of the discipline. Of course, there is the user beloved of entrepreneurs—the consumer.

The fact that the user is so diverse makes it impossible to use the term without specifying its context. One talks of the user in the singular, but one cannot study millions of individuals except as subclasses of a larger whole. It requires classification of user types, some of which are quite obvious, others less so: the male, the female, the normal adult (who is the most obscure of all because he or she lacks other distinguishing features), the child, the elderly, the disabled, the retarded. It is possible to categorize the human by age and sex, capability (e.g., level of intelligence), economic resources (income), and interests (e.g., deep-sea fishing or automobiles). Gross socioeconomic categories are useful primarily in marketing; more individualistic categories are more useful in design, although, as of the present, only potentially.

Paradoxically, despite the importance of the user, we know comparatively little about him or her in relation to how we design. Most of what we know involves physiological threshold limits (e.g., strength, visual and auditory capabilities, etc.). We know almost nothing about the user in relation to the varied facets of technology. We employ users as test subjects in rapid programming (RP) and usability testing (UT), but we do not inquire about them either as individuals or groups. We are only concerned about their performances and preferences in relation to the device we are designing. Indeed, most of the HF literature about users is only concerned with how to elicit information from them to aid design.

Concern for the user as a classifiable type, like women, Blacks, and children, is a recent innovation. The latter part of the 20th century was the period in which concern for previously unconsidered minorities (including the user) blossomed.

Throughout history, merchants have always paid more attention to those who could pay for their goods, but until recently there were only a few of the economically empowered. With the rise of the Industrial Revolution and, more specifically, the 20th century, the increase in disposable income pulled many more into the market. Hence, attention had to be paid to those user characteristics that would cause the consumer to buy. The user, in the garb of the consumer, became ubiquitous with television, followed by the computer, followed by the Internet. The availability of the computer made it possible to draw the user into the equipment design process, and from there it was only a small step for theorists to ask whether human qualities could be designed into the machine. This is where we are at present.

It also follows that we must ask: What do HF design personnel want to do with knowledge about the user? What does the concept of the user actually mean in operational terms? Typically HF has sought to remove technological constraints from the human, so that technology is more comfortable, easier to employ, and so on. The focus on the user here is to learn what impinges negatively on the human and remove those barriers to efficient control of the machine. Much of the interest in the user is as a productive unit—the old Tayloristic concept (Taylor, 1919), but the concept in modern times is that by removing barriers to effective use of a product we create a more satisfied user. This is Taylorism in a more humane guise.

A more compassionate behaviorism seeks to go beyond the removal of work barriers and develop equipment that will respond to human desires and needs by incorporating positive features in the machine. This makes the machine more human. However, this raises the question of which needs and desires should be included and, more important, how we include these qualities in the machine.

Maslow's (1954) basic needs of food, shelter, clothing, and so on are still valid, but in a technological civilization other needs must be conceptualized. When food and shelter have been secured, less basic needs emerge. In a so-called *information age*, one can postulate such needs as curiosity (e.g., the Internet frenzy) and playfulness (e.g., the video game). Beyond this, one might think of the need for intellectual pleasure, but this probably exists only for an elite subset of the total population. For the great majority of the population, there is perhaps aesthetics (the domain of the industrial designer) and an undifferentiated feeling of satisfaction in using machines that work well.

When we think of users, we tend to think of machine operators, but the term is more inclusive. Most people are not operators of heavy duty equipment, but all of us operate household devices. All of us are influenced by technology even if that technology is not a machine.

Few theorists have attempted to incorporate factors in design other than those that represent functionality, but of these, Hudspith (1997) attempted to develop a model for examining design function and user experience together. The model employs dimensions of utility, ceremony, and appeal within which artifacts, design features, user experience, and buyer motivation can be examined. Ceremony is the dimension dealing with human ritual surrounding the use of an artifact. An example is the Japanese tea drinking ceremony. Appeal describes the emotional aspects of user experience, which depends on the user's perception of the artifact, not necessarily its use. Examples can be found in the comparison among such objects as jewelry, perfume, and dress, as compared with garbage bags or soap powder. Prestige also exists in the appeal dimension for those who can afford such machines as Rolls Royce automobiles.

These factors probably only apply to consumer products and more particularly those that involve luxury (which is only partially utilitarian). Curiously, it is surprisingly easy to include ceremony and appeal in design because they rely almost completely on aesthetics.

The user as an abstraction is not useful to the behaviorist and technologist. Much more needs to be known about his or her characteristics with regard to machines. This is a theme for which HF research is ideally suited to provide the answers because, as we intimated previously, almost all HF research as it relates to technology describes how people respond to or interact with equipment. It should be possible to construct a portrait of the user through HF research, but until now this has been done only in the case of one special user class — the aged. We know little or nothing about the great majority of other users.

The study of the user has several dimensions:

1. The user as subject in usability and prototyping tests. Here the focus is on how, and how well, the subject performs relative to a particular equipment or equipment feature. The user as performer is important to all types of hardware/software, military as well as commercial systems.

2. The user as someone who has preferences. Particularly with commercial products, potential users can be asked: Of Designs A, B, and C, which do you prefer? How strongly? Why? Which would you buy?

3. The user as subject matter expert (SME), as an information provider. Design personnel must often rely on the SME because they cannot in a relatively short time develop an expertise gained by years of working on a special type of system. This is particularly true when tasks to be performed are covert or cognitively demanding and the user must be asked to describe his or her mental processes. The SME can be a direct contributor to design particularly if the new design is to automate the SME's activity.

The remainder of the chapter is broken down into the following categories: (a) studies of the user in general, (b) studies of the elderly, and (c) studies of the disabled. In each of these (particularly the latter two), attention is paid to any design guidelines that stem from the available research.

The material in this chapter is based in large part on a review of the relevant papers in the *HFES Proceedings* from 1990 to 1998.

WHAT DO WE WANT TO KNOW ABOUT THE USER?

Data describing cognitive capabilities and actual operation of hardware and software elements are most earnestly desired. Preference data can be secured by testing a prototype, anthropometry by physical measurements, but what is most needed are generic data about performance in relation to equipment and software characteristics.

Problems arise during design that may require this type of information; when they do, the data are often not available in some readily accessible form. A hypothetical question might be: How long does it take an adult male to screw a light bulb into a socket? One would prefer to have these kinds of data in some form of archive, rather than setting up a specific (although a simple) test situation to gather the data.

Before one can gather such data, two types of taxonomy must be developed. A taxonomy is simply a classification scheme that enables one to segregate data into various classes or categories. The first taxonomy needed is one of human functions; this is already available from Berliner, Angell, and Shearer (1964) and Fleishman and Quaintance (1984).

The other taxonomy is of hardware and software elements (e.g., menus in alphanumeric form) and their characteristics (e.g., menu length and depth). This type of taxonomy should not pose much difficulty; one can examine catalogues for controls and displays, software manuals, and HF literature for software characteristics. A primitive taxonomy of controls and displays was developed by Munger, Smith, and Payne (1962) in the course of producing quantitative data describing error rates and response times for each control and display. Chapter 6 included a software taxonomy by Williges and Williges (1984), but a more inclusive, all-technology taxonomy is needed.

How would one wish to use such data? In two ways: First, to differentiate and choose among various controls and displays and software elements (the one that has the lowest error rate and quickest response time would get the nod). Second, to predict operator performance with the selected controls/displays and software elements. The data Munger et al. (1962) collected in their Data Store used error rate and response time as the measures. They collected their data (164 items) from some 2,000 or more HF articles they reviewed. Using the HF literature is more questionable than testing a scientifically selected human sample, but it may be cheaper.

The *iffy* thing about the data that could be collected is that performance context often affects the performance (Moray, 1994). With software in particular, this might be an important variable, and care would have to be taken to develop a neutral (nonbiasing) test context.

Idiosyncratic factors would have to be taken into account—in particular, amount of experience with or level of skill in operating various devices. One of the most common contrasts in the HF literature is that between the novice and skilled operator.

Of course, some data can be secured by systematically examining the HF literature (and data collection should certainly involve this). However, because this involves reviewing someone else's research (and the way it is described verbally and quantitatively), special criteria would have to be developed to select an acceptable research report (e.g., comprehensiveness of description, relevance to the taxonomic characteristics, adequacy of the study, etc.). Meister (1999) provided a description of the processes involved.

Measurement in relation to the individual elements of the hardware and software is, of course, insufficient. It is also necessary to collect performance data with a full-scale device that includes those elements. For example, one might wish to determine the effect of varying the scale intervals of linear displays and then test the subject in perceiving and interpreting various values in the scale.

Something should also be said about measurement of cognitive capabilities. It is unlikely that past HF literature is useful in this regard because, until recently, not many studies were performed in this area. Cognitive measurement requires the development of special tests that should involve some consideration of the cognitive functions required by software, but that would be quite separate from any commercially available software system.

Just as we have anthropometric tables that designers can use in sizing habitations, so it should be possible to develop performance tables for types of hardware/software and types of personnel. All that the effort requires is some interest on the part of a funding agency and the money.

THE USER AND THE COST–BENEFIT RATIO

It is a misapprehension to think of the user as someone who is only acted on by technology, as someone who must use technology or be left in the rear of the parade. How often individuals are told, "How can you live without having a telephone answering machine, a VCR, or a word processor, or without being connected to the Internet?"

Even in a highly sophisticated technological civilization like that of industrialized nations, the individual or even a group of users has a choice. If one is not literally compelled to adopt a technological innovation, one can opt out of the user role. It is possible that millions do. Leaving aside the homeless and impoverished, many of whom have no choice but to avoid much technology, there are still many older adults who foreswear aspects of technology that they fear or distrust.

Even when one is physically required to deal with technology, one can opt out of it partially by accepting certain aspects of that technology and rejecting (ignoring) others. The most illustrative case is that of the military, who, one would assume, would be forced by the nature of their work to use the technology provided them. However, history provides many examples of the contrary.

The first example is somewhat remote in time. In the 1960s/1970s, when the first advanced sonar (e.g., SQS-20) devices were designed, it was found that of the N modes of operation available, the sonar operator tended to use only two or three. All the other capabilities of the device were ignored.

More recently, it has come to our attention that the U.S. Army has attempted to upgrade the M1 and M1A2 Abrams main battle tank by providing a digital electronic command/control work station within the tank. Unfortunately, the increase in information provided by the work station did not improve crew performance in field tests; tankers reported that too much information was being presented.

That is the problem with information as the medium of use in advanced computer systems: Information is not monolithic; there can be too much or too little information. Noncomputerized systems did not suffer from this problem or at least the problem of too much information. The need to supply behavioral resources to deal with information is a side effect of our new technology.

In any event, the result was that tank personnel ignored the new work station capabilities (study described in a proposal to the 2000 HFES annual meeting). If the designer gives the user the opportunity to accept or reject a new technology, the user is just as likely to ignore that technology if he or she sees the human performance cost–benefits received ratio to be negative. Zipf's (1965) law of least effort still functions triumphantly.

Moreover, if the designer makes it impossible to avoid using the undesirable technology, it is likely that excessive stress and errors will result.

Another study involving the army's Theater High Altitude Air Defense (THAAD) radar displays (proposal to the 2000 HFES annual meeting) suggested that when user desires are uncritically incorporated into technology, operator performance may be reduced as compared with a design in which HF specialists employ user desires only as resource material for the new design. In a field test, the user-centered design was significantly inferior to a user-centered, human-factored design.

The user will consciously or unconsciously calculate a cost–benefit ratio in terms of the amount of effort (physical and mental) that must be expended to secure some desired consequence of the new technology. If the benefit is not significantly greater than the cost, the user will reject the new technology (not in toto, but in terms of selecting those aspects the user will use and discarding others to which the user is negative).

If the behavioral specialist can make a stab at estimating that ratio for the anticipated user, a cheaper but more effective design can be developed.

The trick, of course, is to have data indicating how much effort the user will expend for a given technological result. These kinds of data are hard to generalize as a result of general research, but they can be produced as a result of prototype and usability testing during development.

DESIGNING FOR THE GENERAL USER

Methodological Factors

The bulk of recent HF research directed at developing design principles has, curiously enough, not been concerned with design at all, but rather with the development of methods to explore, explicate, and understand the cognitive complexities of the user. Inevitably, this leads us to cognitive task analysis (CTA) and cognitive engineering (CE).

The basic problem is to translate cognitive task data into objective behavioral task requirements. As Hale and Rowe (1998) suggested, the translation process has not been very successful. Some engineers (e.g., Cooper, 1995) feel that the methods used by HF design specialists may be inappropriate to system design — that these methods are more relevant to evaluation.

The symposium Hale and Rowe chaired examined four approaches to the problem, one of which, EID, has been described previously in this book. The other three approaches are: decision requirements tables (Morphew, Thordsen, & Klein, 1998), stemming from the work on naturalistic decision making (Klein, 1993); object view and interaction design (OVID; Isensee, Roberts, Berry, & Mullaly, 1998); and procedural networks (Gillian & Cooke, 1998).

The theorists of EID argue that emphasis should be placed on analysis of the affordances and constraints affecting human–work interaction. For example, EID designers working an aviation problem would concentrate on the significance of altitude relative to flight objects, not simply on the altitude information by itself. OVID focuses on objects as depicted in computer graphics (mail, mailboxes, and users in a communications system and related functions; e.g., reading, sending mail). The procedural network approach emphasizes task data expressed as cognitive behavioral sequences.

The proof of the pudding is in how well each approach transforms data into design. Hale and Rowe said that EID is silent on this point. OVID explicitly claims that their approach will work. The decision requirement technique assumes that if complex cognitive aspects requiring decision making are pointed out to designers, the latter will take the necessary design steps. The procedural networks approach suggests that engineering specifications can be derived from the PATHFINDER network (Schvaneveldt, Durso, & Dearholt, 1985), which are constructed as part of the network analysis.

The genesis of all this competitive theorizing is the increase in system complexity, and this, of course, is fueled by the increasing development of complex cognitive systems. If Taylorism (1919) was the first-generation method of describing and understanding what the operator was supposed to do (physically and perceptually), the present generation concentrates on cognitive demands, and particularly on the decision making required of the operator. Whereas first-generation systems allowed the HF specialist to observe the operation of controls and displays, much cognitive activity is covert and therefore requires some way of making the covert activity transparent to an observer.

It was possible with earlier computer systems to develop guidelines of the Smith and Mosier (1986) and Williges et al. (1987) types. Although these guidelines did not lead to design as such, they could be used as evaluational tools.

The present research on CTA and cognitive mapping is at a more embryonic stage. In the framework described by Potter, Roth, Woods, and Elm (1998), CTA should be thought of not as a single methodology, but as a *breadbasket* of techniques that are to be applied as the situation suggests. Their approach conceptualizes CTA as a process rather than a method:

> CTA is an iterative, bootstrapping process focused on understanding the domain (mapping the cognitive demands of the field of practice) and the practitioners' modeling expertise and cognitive strategies. . . . CTA extends into the design prototype stage; supposedly cognitive TA outputs become virtual hypotheses for design practice which, when tested, suggest refinement. (Potter et al., 1998)

So much is not much different from what ordinarily goes on in design (particularly the use of prototypes), but has the virtue of connecting CTA directly with design products.

The illustrations provided by the theorists suggest the variety of techniques they subsume under CTA: semantic mapping, ethnographic/observational investigations, critical incident technique, critical decision methods, structured interviews, means–ends analysis, and task and workflow monitoring. One uses whichever technique seems to be most appropriate for the situation under investigation.

When one analyzes the verbalism in Potter et al. (and in the other CTA theorists), it appears that we are dealing with traditional phenomena in the analytic design phase (TA, which emphasizes cognitive aspects of behavior, interviews, critical incidents, etc.) and the prototype testing phase (which is more recent than TA, but now sufficiently enshrined at the heart of the design process to be considered traditional).

It is entirely possible that what we are really dealing with is what we can call *retrograde design*. It becomes necessary to understand how to translate the results of CTA into CE before one can develop an appropriate CTA methodology. If a design engineer says to the HF specialist, "I can use the results of your CTA only if it can be translated into a flow diagram," this requirement immediately sets at least one of the parameters of the CTA methodology. In a perverse way, the analyst must anticipate the translation of the behavioral data into hardware specifics before appropriate behavioral data can be gathered. In the absence of such retrograde design, the CTA data gathered may be interesting but entirely inappropriate to the demands of the interface design. The problem is similar to the one found in developing an experiment; if the appropriate statistical analysis is not specified in advance, it is likely that inappropriate and nonrevealing measures will be selected for the study.

Idiosyncratic Factors in Design and Operation

We mentioned previously that user-centered design aims to make design responsive to positive user traits, such as the impulse to experiment, curiosity, intuition, cognitive integration of stimuli, and amusement. Because we are talking about informational rather than procedural systems, it is theoretically possible to incorporate design features that will elicit such responses because these responses relate most directly to information utilization.

The heavy emphasis on displays in computerized systems (graphic user interfaces [GUI]) takes advantage of these user characteristics because these displays require perceptual/cognitive integration capabilities.

The possibility that the information stimulus can be interpreted in more than one way is what makes the user more important to software than to hardware design. In the latter, multiple interpretations are not possible, only deviations from the predetermined procedures, which we call *errors*. Where there are multiple ways to interpret information, and particularly where the operator has multiple options to proceed, the traditional definition of error (a deviation from procedure) no long applies. However, this is what gives software/information its increased potential. Flexibility is potential. The opportunity for the designer is to take advantage of that flexibility in the design. However, this becomes a problem when the designer does not know how to apply that flexibility.

At the same time, there are cognitive limitations that are inherent in flexibility and that must be avoided in design. The most important of these is information overload (limitations in terms of the number of stimuli that can be assimilated by the operator (Miller, 1956; Schroder, Driver, & Streufert, 1967). The overload results from the multiple possible interpretations of stimuli that flexibility permits. These multiple possible inter-pretations demand more intensive analysis by the operator, which may lead to overload. The designer's problem is to be able to predict (quantita-tively, preferably, or qualitatively, if nothing better can be achieved) just how much information is being processed through the system. Generaliza-tions to the effect that excessive information will degrade performance are simply not good enough; information utilization must be at least scalable.

What is needed are equations representing types and amounts of infor-mation that will produce certain performance levels when presented by displays. Once the amount of information can be predicted, it is possible to work one's way around the anticipated difficulty by isolating informa-tion into separate streams, assigning another operator, or combining in-formation into integrated predictor displays.

Amount of information interacts with the speed with which informa-tion is presented and the number of changes in that stimulus information not only during any one operating cycle, but also in response to the exter-nal situation, as in emergencies. The negative effects of information load are exacerbated by speed and change, just as low temperatures are low-ered even more by wind conditions — the so-called *wind chill effect*. Thus, there may be an information chilling effect, but we will never know unless systematic studies of information processing related to system character-istics are performed.

The key to including positive factors in design may be to introduce greater flexibility, the opportunity for the user to select alternative options with regard to the nature and amount of information that can be analyzed in various ways. The system should be adaptable to the user's intellectual resources. If the designer could vary the amount and speed with which in-formation reaches the operator, this would encourage the user's tendency to experiment, which is directly linked to his or her curiosity. Indeed, some of the attraction of the Internet is directly associated with the flexi-bility the user has in *surfing the net*. This capability is also tied in with amusement (flexibility → experimentation = amusement).

All of this may be somewhat idealistic. Systems are designed to do cer-tain things, and these may require certain types and amounts of informa-tion presented at certain speeds. If the system is relatively inflexible, so be it. However, information inherently has the capability of being flexible, and the designer should look for opportunities to increase that flexibility.

Flexibility also ties in with experience. It is not merely that experience improves skills, which is well known, but that, in connection with infor-mation processing, experience leads to alternative strategies. For example, Matthews and McFadden (1993) found that experienced sonar operators used a recognition-primed approach that is characteristic of time-critical decision making. They formulated a hypothesis about the target based on previous experience with similar patterns. They then analyzed specific aspects of the pattern to evaluate their hypothesis. Operators lacking that experience would be less likely to develop and test a hypothesis.

Where computer-generated decision aids are involved, the attitude of the operator toward the computer may affect cognitive processes. In the case of a decision aid for flight planning, some planners failed to understand the computer's limitations (Smith, McCoy, & Layton, 1993). They may fail to use critical computer-generated information at appropriate times. Too much information may be generated for the user to accept. The tendency to adopt an overly positive attitude toward computers and the information they present may lead to an unwarranted reliance on them for critical decision making. However, very early 1950s studies of decision aids for submarine commanders revealed that some personnel were unwilling to waive their own judgment in plotting an attack course in favor of the computer's recommendations.

Not much attention has been given to individual differences in operation of computerized systems. Lee and Moray (1992) studied a simulated process control plant and found that individual differences in monitoring patterns (symptomatology) during normal plant operations corresponded to differences in ability to mitigate the effects of faults in the system. The study of individual differences in performance could lead to the discovery of stable stereotypes in perceptual/cognitive activity.

USABILITY TESTING

The questions that animate this discussion are: What does research tell us about how usability testing (UT) should be conducted? How does UT differ from prototyping? Finally, does UT tell us anything about users?

Much of the usability testing literature deals with the mechanisms of how such tests should be conducted. For example, Catani and Biers (1998) explored the fidelity of prototypes of a library computer system offered to experimental subjects in the form of paper, screen shots, and an actual computer system. They found no difference in the number and severity of problems encountered or in the evaluation of the product. Usability specialists and users differed in the incidence and commonality of the problems discovered, and the professionals differed among themselves in their ratings of the problem severity. The problems found in actual testing differed from those found by subjective evaluation.

This study points out that there is also a strong element of subjectivity in the usability test, particularly in what has been termed *heuristic evaluation*. This, as described by Nielsen and Molich (1990), is an informal method in which one or more professionals comment on an interface design.

In a heuristic evaluation, there are neither actual users nor actual task performance. Consequently, heuristic evaluation may lead to results that differ from those of actual performance (Desurive, Kondziela, & Attwood, 1992; Jeffries, Miller, Wharton, & Uyeda, 1991).

Heuristic evaluation is the same traditional, informal evaluation of a design product by one or more designers with or without published guidelines as an evaluation tool. This is another example of the way in which traditional methods are dusted off and recycled, as if they were novelties.

Nielsen (1990) had two groups of professionals evaluate either a computer- or paper-based prototype of an existing system. For the same number of problems (50) found in both prototypes, different types of problems were identified depending on the type of prototype. The computer-based prototype identified more major usability problems; the paper prototype, more minor ones.

Another study (Fu, Salvendy, & Turley, 1998) compared heuristic evaluation and usability testing and found that for different levels of user performance, different types of information are processed, and users make different types of errors. These researchers classified errors into three types following Rasmussen (1983): skill-based, rule-based, and knowledge-based. HF specialists were more effective in identifying skill- and rule-based problems, and users were more effective in identifying knowledge-based problems.

The various evaluational methods that can be used in usability testing have been classified by Whitefield, Wilson, and Dowell (1991) based on how the user and computer components of the system are presented in the evaluation. Each of these two components can be real or representational. For the computer, *real* meant the actual computer software or a prototype. A representational computer might be the specifications of the device or the user's mental model of the computer, as described by questionnaire or interview results. For the users, *real* meant actual users or their approximation (e.g., students). A representational user is a description or model of the user as represented by experts performing a heuristic evaluation.

Based on the preceding classification system, these researchers divided the methods into four types: analytic methods, specialist reports, user reports, and observational methods. Heuristic evaluation involves a representational user and a real computer. The evaluators examined the interface in accordance with recognized usability principles. Typical of these is what Nielsen and Molich (1990) proposed: The interface should (a) use simple and natural dialogue; (b) speak the user's language; (c) minimize the user's memory load; (d) be consistent; (e) provide feedback; (f) provide clearly marked exits; (g) provide short cuts; (h) provide good error messages; (i) prevent error; and (j) provide help and documentation.

None of these heuristics is operationally defined except perhaps in the expert's mind, none is quantified, all are intensely subjective, and all represent goals rather than measurable criteria. The subjectivity inherent in evaluation using such guiding principles guarantees that experts performing the evaluation will disagree. One might as well add Mom's apple pie to the list of principles; it will be just as effective as the preceding.

User testing involves a real system and a real user interacting with a real device. The testing is conducted in a usability laboratory (Nielsen, 1994, 1997; Scerbo, 1995), which presumably provides more control and less subjectivity. User performance measures can be collected, as well as think-aloud protocols, and scenario-based testing can be employed.

Usability testing (UT) is part of and the end process of user design. In the context of performance measurement as a whole, it has similarities with experimental research, with performance testing of the operational system (OST) type and with prototype testing. The purpose of the control that is exercised in UT in a laboratory environment is not the same as that of experimental research; the latter tries to discover which variables influence the performance of interest, and UT is concerned only to ensure that biases which might distort the data do not occur. UT may or may not be performed as it relates to a complete system; it also often includes prototypes and simulation. When performed properly, OST always includes the full operational system. UT is closest to prototype testing, but prototype testing may occur outside the laboratory.

In all these variations of performance testing, the following factors must be considered, sometimes concurrently, sometimes individually:

1. The test must be planned; it is often valuable to perform a pretest to ensure that all will go well in the final test.
2. The purpose of the test and its implications for data must be decided before testing.
3. The attributes of the device that one is testing (e.g., flexibility, multiple user options) must be specified.
4. The measures to be used in collecting data must be derived from the attributes of interest.
5. The criteria (e.g., number of errors, response speed, preferences) that indicate whether performance is adequate or inadequate must be specified before testing.
6. Subjects for the UT must be described: who they will be (other specialists, the general population, students) and their number.
7. The tasks that subjects will perform in the test and what subjects are expected to do to provide data.

8. The role the evaluators or data collectors will play in the test must be decided on: Do they simply stand around and supervise, provide necessary information if subjects ask for this, or act as observers — all this cannot be simply inferred by test participants.

9. The structure of the test must be specified (e.g., number and duration of test trials, number of test repetitions).

10. The same applies to the nature of the stimuli to be presented to subjects (e.g., a prototype device, a simulation, a scenario, a set of photos, etc.).

11. The test instrumentation to be used (e.g., video/audio recorders; eye trackers; key stroke logging) must all be listed.

12. How test data will be collected (ties in with 11) must be described.

13. A plan for analyzing the data must be developed in advance (e.g., statistical analysis of performance, analysis of subjective data).

Like most other performance tests, UT will tell us nothing about users in general, nor very much of them in particular. UT will reveal preferences if subjects are asked for these, and consistent preferences over a large enough number of subjects might be generalized to users as a whole; but UT is intensely particular. The intent of the UT is not research on users, but merely to use them as subjects akin to running rats in a maze. Any behaviors of users, even in the UT, can only be interpreted relative to the test. Also companies regard their UT as proprietary information and often strenuously resist publication of UT data.

However, there is some potential in UT for learning about users as test subjects (e.g., do they show evidence of learning any skills from the test? Are they flustered by the test situation and the test experience? Do they prefer certain kinds of equipment features, e.g., keyboards as entry devices as opposed to clicking a mouse at a menu?). The UT evaluators will certainly learn something about users of the device they are developing; that is, after all, the purpose of the UT.

DESIGN FOR THE ELDERLY

Introduction

It should hardly be necessary to explain why we include a section on design for the elderly in a chapter on users. The population in industrialized countries is becoming progressively older, and the elderly are users. They buy and use equipment, they interact with technology, and their ergonomic needs require satisfaction.

The stigmata of the elderly are, in most cases, not catastrophic losses unless a disease like Alzheimer's or Parkinson's is involved. Most commonly, there is a progressive fading away and a gradual reduction of capability, which are often added to bizarre symptoms like allergies or a greater sensitivity to cold temperatures, which are nuisances rather than life threatening.

Losses may occur in the following faculties:

1. Sensory/perceptual: poor eyesight and loss of hearing;

2. Physical: locomotion difficulties, weakness, trembling, excessive fatigue;

3. Conceptual: loss of short-term memory, slowing of responses to information that appears quickly, difficulty in interpreting masses of information, difficulty in learning new processes; and

4. Personality: gradual withdrawal from external society, suspiciousness, unwillingness to try new objects and phenomena, mood changes, depression, loss of self-confidence. Personality factors are largely outside the purview of ergonomics, but certainly interact with it (e.g., loss of self-confidence leading to an unwillingness to try new technology).

Of course, not every aged individual displays these symptoms; many people live into their 90s without manifesting any of these losses, but they are sufficiently familiar that they have become a stereotype of the aged.

The symptoms listed earlier are essentially deficiencies that can be treated by rectifying the deficiency (developing compensatory mechanisms, technology, or prostheses). Although age is not a disease, from an ergonomics standpoint, it can be treated or ameliorated if the deficiency can be identified. A deficiency (e.g., loss of a limb) inevitably suggests a compensation for the deficiency.

As a consequence, one needs no elaborate theory or model to describe what is happening or what needs to be done. The aged provide their own compensatory models by adopting behaviors that seek to deal with their deficiencies. For example, if one becomes abnormally sensitive to cold, one wears more sweaters as the winter approaches.

If there is a tendency to slip and fall, one can install grab bars in the bathroom. The kind of theory making that we have observed in trying to understand the behavior of the normal, younger adult is necessary because their problems are often obscure and difficult to identify. However, the symptomatology of the elderly usually dominates these more obscure normal behaviors.

What we have talked about so far is remediation. Vanderheiden (1997), whose discussion of design for disability is very comprehensive, emphasized something called *universal design*. He defined the term as the practice of designing products or environments that can be effectively and efficiently used by people with a wide range of abilities operating in a wide range of situations. This includes people with no limitations as well as [those functionally limited]. (p. 2014)

Universal design principles are presently under development, but the final version is not available for examination. However, Vanderheiden did provide a list of *pro tem* design principles that, on examination, seem to be similar to those that guide human engineering in general, although with special concern for those with functional limitations.

Vanderheiden's list, which is abbreviated in Table 7.1, includes seven principles with a number of guidelines for each principle.

Any product designed to these principles would be effective for everyone, functionally limited or not. One principle — accommodation of preferences and abilities — would require further empirical research to determine what these preferences are. The equitable-use principle is a goal, not a design methodology. Simplicity of use is, of course, a basic principle of all good human engineering, but how to define it in any given design situation is very difficult.

TABLE 7.1
Principles of Universal Design

Principle	Description
1. *Simple and intuitive use*	Product use is easy to understand by reducing complexity, making design consistent with user expectations, making information clear, and providing for feedback.
2. *Equitable use*	The design does not disadvantage any user group by providing equivalent access and efficiency.
3. *Perceptible information*	The design communicates necessary information effectively regardless of ambient circumstances and user sensory limitations.
4. *Tolerance for error*	Design minimizes consequences of accidents by locating components to minimize errors and hazards, providing error warnings, and making inputs reversible.
5. *Accommodation of preferences and abilities*	The design accommodates a wide range of preferences and abilities by minimizing the need for dexterity, facilitating user accuracy, and providing adaptability to user pacing.
6. *Low physical effort*	Design can be used comfortably and with minimum fatigue by minimizing sustained effort, maintaining a neutral body position, and requiring minimal repetitions.
7. *Space for approach and use*	Design provides appropriate space regardless of the user's body size or mobility. Factors involved are: clear light line of sight, comfortable reach, and accommodation of prostheses.

Note.   Adapted from Vanderheiden (1997).

Although one can laud the concept of universal design, it must overcome certain practical difficulties. The first is the translation of design goals into design methods. For example, the reduction of complexity (simple and intuitive use) is a basic principle; in theory, a design should not be any more complex than it needs to be. How is the designer to determine how much complexity is required to achieve functionality? Second, realistically, universal design will be expensive, and entrepreneurs react violently to added cost unless someone else pays for it. It could be said that more effective design will encourage more users to purchase a product, but few entrepreneurs think in these terms.

Research on the Elderly

There is an obvious paradigm in research on the ergonomics of the aged. This involves selecting some equipment or task and exposing subjects of different ages to that equipment or task. The performance differences between the two or more sets of subjects (the aged vs. the young) are then interpreted in terms of underlying behavioral processes. This paradigm, in effect, says that performance of the elderly, which deviates from that of the younger cohort, indicates a deficiency that requires some corrective action. This places a premium on youth, which only a youth-centered civilization could accept. From a purely conceptual standpoint, therefore, ergonomics research on the aged presents few difficulties. The interpretation of what is found may be more difficult.

One of the questions the researcher must ask is whether the task selected for implementing the paradigm (young vs. old) is appropriate for the young versus old test. Is any task meaningful for this comparison? Not where the task is performed only by the young or the old. Thus, one might not compare young and old on motorcycling ability, skiing, or surfing, which are activities engaged in largely, if not solely, by the young. The task must be performed by both cohorts and with equal frequency to be meaningful. An example is operation of the automated teller machine (ATM; Jamieson & Rogers, 1998).

Research in the form of a controlled experiment is not the only means to discover information relative to aging. Rogers, Meyer, Walker, and Fisk (1998) used focus group interviews to assess living constraints of individuals ages 65 to 88. These constraints were coded as to whether they were remediable by training, redesign, or a combination of the two. Fisk (1999) pointed out that more than half the problems reported could be helped by HF interventions, as were almost all the cognitive problems.

Fisk suggested that regardless of whether they like it, elderly users continue to encounter new technology that may stress them. Those with more self-confidence wish to conquer these technologies; those with less self-confidence tend to reject them by isolating themselves from the technology.

There have been a number of efforts, both in the United States and abroad, to get the elderly interested in technology, particularly in computers. Czaja (1996, 1997b); Czaja, Clark, Weber, and Nachbar (1990); and Rogers (1999) examined technology training for older adults. Mannova (1999) actually enrolled seniors in an information technology course; she found that, once the initial fear of incompetence was overcome, there was great enthusiasm. Seniors are apparently not inherently technophobic or disenfranchised with regard to technology. Access to technology is not only a matter of physical and economic capability, but to an equal extent involves entering new sociocultural contexts (Mynatt, Adler, Ito, Linde, & O'Day, 1999).

There have even been attempts to involve the elderly in relationships with designers and in research and development (Hofmeester, Dunne, Carver, Susani, & Pacenti, 1999; see also Ellis & Cochran, 1999).

As one would expect with an HF research that has always emphasized automotive themes, much HF research with the elderly has emphasized driving (e.g., Barr & Eberhard, 1991; Staplin & Fisk, 1991).

Many of the solutions to problems can easily be achieved if we are sensitive to them. For example, many of the aged refuse to use ATMs. A survey conducted in Memphis and Atlanta by Rogers, Fisk, Mead, Walker, and Cabrera (1996) and Rogers and Fisk (1997) of 1,500 adults revealed that the percentage of use by younger adults (ages 18–34) was 86% versus 33% for those ages 65 and older. Design of and lack of training in ATM use were the primary reasons for refusal to use the machines. Those who designed these machines felt that their use was almost intuitive, and only 15% of banks had pamphlets with instructions. Mead and Fisk (1998) found that systematic training increased the percentage of use by the elderly to 80%.

This type of research suggests a number of design principles that can be useful to all, not merely the elderly. A major principle is to embed training in the design of an equipment like the ATM. This is not a new and startling principle; embedded training has been around since the 1960s.

Because computers are here to stay, design must deal intelligently with menus, as these are presently the primary means by which one navigates through a computer system. In connection with computers, Walker, Millians, and Worden (1996), and Walker, Philbin, and Fisk (1997) found that, because of poorer motor performance, older adults have more difficulty using a mouse to position the cursor, which can be a major obstacle to computer use by the elderly. Fisk (1999) suggested that the solution to this problem might be to make software changes in the gain and acceleration profiles that translate mouse movement into cursor movement. Some of these problems can be overtaken by advances in software and hardware capabilities (e.g., there already exists an automatic hooking algorithm in which the cursor [hook] is attracted to the nearest target on the screen). As one moves the mouse, the cursor will leap to the next nearest target. A double click activates the target (icon) file or program.

In any event, Worden, Walker, Bharat, and Hudson (1997) have shown that, with proper interface design (e.g., an area cursor with sticky icons and adaptive gain control), older adult performance can be improved by 40% without training; such an interface also improved the performance of younger adults by 20%.

Skill-acquisition studies have found that, with extensive practice, performance becomes highly specific to incidental features of the training environment. These are features or relationships that act as cues in task performance. In a sense, they become like helpful icons or landmarks in a navigation task. Two studies by Meyer and Fisk (1998) examined the role of these in training of a visual search skill and logical reasoning. It appears that older adults are less sensitive to this information than younger ones. From a design standpoint, it appears that adding such features, which consistently relate to correct actions, can improve the performance of the aged.

Lowered performance for the elderly may be due to reduced information-processing speed or time-sharing ability, as found by Kramer, Larish, and Strayer (1995) and Salthouse, Fristoe, Lineweaver, and Coon (1995). The ability to process rapidly changing stimuli can be significantly reduced in the older adult. Nevertheless, such deficits may be an artifact of a type of test that emphasizes speed; many older adults will not expose themselves to rapidly changing stimuli unless these are unavoidable.

There has been an intensification of interest in the ergonomics problems of the aged, as represented by research published in the journal *Human Factors*: special issues in 1981 (Fozzard, 1981) and 1990 (Czaja, 1990), as well as two special issues devoted to elderly drivers (Barr & Eberhard, 1991; Eberhard & Barr, 1992). Rogers, Meyer, Walker, and Fisk (1998) used focus groups to explore the limitations of daily living for the elderly. A focus group is a group interview, typically from 6 to 10 individuals, who are brought together to discuss their problems. The University of Birmingham's (England) Institute of Gerontology has a volunteer group of 4,000 to 5,000 people organized into something called "A Thousand Elders," who make themselves available for interviews, questionnaires, prototype evaluations, and so on (Hayak, 1995). In the United States, one can recruit the elderly at senior centers, retirement communities, and so on. The center of interest in such focus groups is the constraints and difficulties experienced by the elderly; one of the interesting things discussed by Rogers et al. (1998) is the response to their constraints and difficulties: 57% ceased their activity in response to task difficulties, 22% persevered, and 19% attempted to improve their performance. New technology in the home (e.g., VCRs, telephone menus, home security systems, answering machines, cameras, microwave ovens) presented continuing challenges.

It is very likely that many more of the aged would accept new technology if appropriate training were provided. The lack of adequate training is a problem that faces not only the elderly; the younger, *normal* adult is often provided inadequate training in the use of new technology. The lack of adequate instructions for operating PCs has been scandalous for many years. Simple, understandable instructions require no redesign, merely the will on the part of engineering and management to do better.

One characteristic of ergonomics research on older adults is that, although it tells us that the elderly perform less effectively than the younger, it presents few, if any, recommendations for improving the situation. An entire issue of the journal *Human Factors* dealing with driving skills of the aged (Barr & Eberhard, 1991) leaves us with no recommendations except that enhanced left turning signals at intersections would improve the driving performance of both younger and older drivers (Staplin & Fisk, 1991).

We see in the next section that it is possible to make specific design recommendations for motoric disabilities, which many of the aged possess. For perceptual/cognitive problems, the ergonomics literature seems to have few, if any, concrete recommendations except enhanced training for ATM users and larger and simpler warnings on medication labels. It is impossible to avoid the impression that most of the research on the aged leads nowhere except to themes and models of why the deficits in aging occur.

The difficulties that the aged experience with new technology may disappear as the older cohort dies out. Youngsters conditioned to new technology may do better as they grow older unless still newer technologcal developments, in turn, leave them at a disadvantage.

DESIGN FOR THE DISABLED

Introduction

Dealing with physical problems takes precedence over solving cognitive problems. If it is difficult to get in or out of a bathtub, problems with ATMs and VCRs seem less important.

We include among the disabled those elderly who possess perceptual and motoric disabilities that make it difficult for them to perform ordinary living activities; those who are physically handicapped (of any age), such as paraplegics; and those who are mentally retarded. The amount of ergonomics literature related to the last two categories is minuscule in comparison with that dealing with the elderly, but we review it nonetheless. For the elderly who are physically handicapped, specific design recommendations for common household equipment are reviewed. These recommendations do not necessarily come from ergonomics specialists, although obviously the recommendations are of an ergonomics nature. The sparsity of our disability literature may be because we confined ourselves to the HFES *Proceedings*; there may well be a larger literature that is medically oriented. However, because we are dealing with an ergonomics audience, we have chosen not to include that larger literature if it exists.

If the ergonomics literature on the disabled is sparse, it is because HF researchers have chosen (consciously or not) to describe the normal population. This is entirely rational; HF researchers are almost entirely unimpaired.

Within the HFES literature, we have been able to find only eight studies of the disabled over the last 10 years. The most recent study is by Lee (1995), which dealt with access to computers by the visually impaired (meaning, essentially, blind). The Lee study examined the possibility of using the haptic modality for access to computer systems. The study compared sighted and unsighted subjects (the common paradigm) in using raised icons. There were no differences between sighted and unsighted people in making this distinction.

A study was also performed to design an electronic memory device to enhance medication compliance in the elderly. The average elder takes 4 to 5 prescription medicines daily, while one in nine takes as many as 10 or more (Ausello, Lamy, & Gondek, 1990). The little research on the medication compliance and memory enhancement issues (e.g., Leirer, Morrow, Tanke, & Panante, 1991) suggests that memory aids for those elderly who commonly experience memory problems would be useful.

In the Ausello et al. study, design concepts for memory devices were developed using a storyboard methodology and presented to a sample of 30 elders. The researchers commented that designing a device that meets specific criteria and does not intimidate the older user is an engineering challenge. Displays that present text must be large enough to read, layout should be uncluttered, contrast should have an acceptable level, dates and times should be presented in a standard form, icons should be unambiguous, tactile and auditory feedback should be provided with each button press, spacing should be generous, buttons should be guarded against accidental activation, and so on. These are attributes and criteria, not specifics, but each one is relatively easy to accomplish, although perhaps not all in combination.

Design for Living

Design for the disabled centers around basic housekeeping activities, such as going to the toilet, washing, and preparing food. Therefore, it does not require complex computer software, which is a pity. If it required computer technology, it would probably have attracted much more technological attention than activities supporting more basic needs.

Keeping oneself clean in safety (the bathroom) and being able to cook for oneself (the kitchen) represent much of what the disabled (including those elderly with physical impairments) desire. The following is modified from Pirkl (1994).

Four attributes are necessary for achieving environmental quality: legibility provides environmental cues that permit the user to orient and direct him or herself; accessibility permits the user to move freely through an environmental setting; adaptability determines the range of adjustment required by the environment; and compatibility permits artifacts and spaces to be amenable to people's functional limitations. These attributes can be summarized in phrases such as safety, comfort, and ease of use—qualities that humanize an otherwise inhumane technology.

Generic Guidelines

Vanderheiden (1997) is particularly good at providing suggestions for alternative design possibilities for the functionally limited, and the reader is urged to consult this reference. The list of recommendations he provided is too lengthy to be included in this chapter. We have made greater use of Pirkl (1994), whose guidelines overlap substantially with those of Vanderheiden.

The following are generic guidelines (Pirkl, 1994) for designing or selecting transgenerational design products:

1. Provide cross-sensory redundant cueing for all alarms, signals, and controls (combine an audio signal with a visual indicator).
2. Offer redundant modes of operation utilizing the next larger set of motor movements (finger to hand, hand to arm, arm to foot).
3. Establish consistent display/motion relationships (left to right and forward/up to increase, backward/down to decrease).
4. Provide definitive feedback cues (control positions [*détentes*] should snap into position).
5. Reduce the complexity of all operations (minimize the number of tasks).
6. Place critical and frequently used controls within easiest reach (cluster controls on basis of priority).

7. Prevent accidental actuation of critical controls (relocate, recess, or provide a guard).
8. Provide adjustable product/user interfaces (horizontal/incline, vertical/incline, raise/lower, push/pull).
9. Design for use by a variety of populations (male/female, young/old, weak/strong, large/small).
10. Design to facilitate physical and cognitive function (encourage user to practice, and improve by making operations easy and enjoyable).
11. Design beyond the basic physical/functional need (enhance the user's independence, self-respect, and quality of life).
12. Compensate for a range of accommodation levels (provide for some exercise through user interaction/participation).
13. Strive to make task movements simple and understandable (clockwise for *on* or *increase*, counterclockwise for *off* or *decrease*).

The preceding guidelines are familiar staples of good human engineering practice. The reason for reiterating them here is that they are particularly important for the elderly/disabled. Younger, normal adults can more easily make the adjustments required if these guidelines are violated; the elderly/disabled have greater difficulty.

The Bathroom

Pirkl (1994) included a special section dealing with the bathroom. It is unfortunate that as one considers design for the elderly, one must deal with primitive functions almost as if one were dealing with the newborn. One must begin with the toilet and the bathroom.

Pirkl had been anticipated by Kira (1976) and has been updated by Hayak (1995). In Kira's words, the average bathroom is

> hopelessly antiquated and inadequate, and still consists of a miscellaneous assortment of oddly sized, unrelated and minimally equipped fixtures, with inadequate storage, counter space, lighting and ventilation. (p. 17)

The standard bathroom environment poses crucial problems for the elderly and mildly handicapped:

> Without adequate provisions for support, using the bathroom may cause slips and falls. Many elderly . . . find it difficult to get up from the toilet. . . . The absence of stabilizing supports in the washbasin area causes problems to people with declining balance. . . . Protrusions and sharp corners can lead to serious injury in case of falls. Fixture controls are often . . . difficult to use,

... for those with arthritic hands. .... Access to the shower/tub unit is not
only inconvenient, but often unsafe. (Pirkl, 1994, p. 23)

Pirkl (1994) went into considerable detail in describing a remodeled
bathroom designed for the elderly/disabled. This could be installed in a
bedroom closet or an adjacent hallway or closet. The location close to the
bedroom eliminates much of the difficulty experienced in locomotion.
Newer homes may have bathrooms located much closer to the master
bedroom.

Such a facility would have support bars and guides related to the toilet
and sink unit. Because the elderly/disabled may need support in moving
from toilet to sink to shower, the bathroom should have a number of grab
poles, including one in the shower unit.

Four guidelines for design of the bathroom storage area are: optimal
visibility and reach for contents, avoidance of storage density, avoidance
of the need for acute bending, and a proper place for support (especially
when the hand is raised).

The Toilet

Much as one might wish to ignore it, the toilet represents one of the most
basic human functions to be considered.

Seat Height.   The dimensions of seat height, steady support, and safe
transfer from a standing position onto and off the seat pose two main design considerations. The most significant difference in seat height requirement is between men and women. However, because more than four
times as many older women than men live alone, and older people of both
genders are shorter than younger people, the suggested seat height is below the common geriatric level. Although it is generally recommended to
elevate the toilet seat by 2 or 3 inches for the elderly because of their difficulties in coping with low seating, it is desirable from a physiological
standpoint to design the seat to encourage a tendency to lean forward and
draw one's feet back to facilitate elimination. This encourages defecation
in a squatting position. Because the aged often suffer from constipation,
and thus must spend much more time on the toilet, a lower seat helps
avoid circulation problems while encouraging a posture to aid the body's
natural mechanisms. The suggested seat height above the floor should be
15¾ inches.

Body Support.   The major problem in designing an appropriate water
closet is to find ways to transfer from the seat to a standing position. The
ledge in front of the lavatory can function as an integral part of the toilet,

supporting the individual to pull him or herself up. In addition, a vertical pole next to the toilet and the flat surface on the sides of the seat facilitate a safe transfer.

Special Features.   The control bar of the flush mechanism should be placed at shoulder height above the toilet to be operated from a seated or standing position. A long and shallow flush tank should be mounted high inside the plumbing wall.

The Lavatory.   The washbasin is the most frequently used fixture in any traditional bathroom. Lavatories are usually mounted too low, so that the body is forced to bend over, causing a tiring posture that cannot be long sustained. The most comfortable working height of the hands for the average individual is at 38 inches, with the water source at 42 inches. The lavatory and its related surfaces should be mounted between 34 and 38 inches.

The proposed lavatory can be integrated into the countertop facing the toilet at a height of 36 inches. The opening of the basin should be 18 inches wide, 15 inches from front to back, and only 4 inches deep. The water source should be 42 inches above the floor level, with the arc reaching well into the center part of the basin.

Shower.   Showering is the most common and safest form of body cleansing for the elderly. However, a common problem is the need for prolonged standing, inadequate clearance for washing, and absence of support.

Because the main orientation of the individual is toward the water source and the shower enclosure is rectangular, the water source and the shower seat should be positioned along the axis of the water stream. The inner dimensions of the shower enclosure should be 40 × 35 inches. The width above the open access shelf (35 inches above the floor) should extend by 4 more inches to 39 inches. A lower barrier 3 inches in height can be provided at the entrance to contain the water of the shower spray. What has been said about support bars and guides may be in the process of being outdated. Doors can be rigged with photoelectric door openers similar to those used in stores with automatic access. Other manually op-erated controls (e.g., pushbuttons) are also available.

Pirkl's treatment of the bathroom has been empirically tested by Hayak (1995). The procedure involved the "Thousand Elders" of the UK and the use of a series of focus groups to provide recommendations. A series of physical prototypes of a bath, washbasin, toilet, and faucets were devel-oped and actually used by individual groups of the elderly. Design changes were made based on successive testings. In general, the design of the bathroom followed Pirkl's recommendations.

DESIGN FOR CHILDREN

There is little in the HF research literature that pertains to children. In reviewing the HFES *Proceedings* from 1990 on, we found only two studies that involved children (McCrary & Williges, 1998; Wise & Wise, 1991). Of course, much research related to the safety of devices used by children (e.g., toys, furniture, playground equipment) is performed, but the Consumer Product Safety researchers in government publish their findings in other than HF journals.

DESIGN FOR THE HANDICAPPED

A few innovative studies have attempted to use computer technology as a means to teach the moderately mentally retarded. Robertson and Hix (1994) developed a three-phase study to examine the capability of this population to use a PC. The term *moderately retarded* must be interpreted. The 12 subjects had a Stanford Binet IQ of between 35 and 50, which we would rather consider as *severely retarded*.

PC use was built around a game designed to teach dressing and shopping practices. The study produced a number of design guidelines:

1. The pace of the stimuli on the screen should be completely under the control of the user;
2. Window boundaries should be locked in place, rather than being manipulable;
3. Active bars need to be 1.5 inches from menus;
4. Strong visual feedback is necessary to comprehend a concept like *putting articles back*;
5. The meaning of icons must be taught;
6. Screen clutter should be avoided; and
7. The computer should be utilized as a shared activity.

Disability also occurs in physical motoric functions, as in quadriplegia. There is significant practical value in assisting physically disabled people to use computers. Casali and Chase (1993) studied 10 persons with hand and arm dysfunction as a result of spinal injury. Their task required subjects to select targets of various sizes and distances using cursors. Such people can team to handle computers, and their performance can be enhanced by embedding simple training exercises in the software.

Patients with ALS (Lou Gehrig's disease), who are eventually reduced to expressing themselves with eyeblinks, can also be assisted by computer technology. In the case described by Murphy and Basili (1993), the patient communicated by an eyeblink sensor that was reflected in a computer screen displaying alphanumerics. What this study revealed is that it is possible to make design modifications that satisfy the special requirements of the physically impaired.

Using common household products is often difficult for people with neuromuscular disorders and arthritis. A number of studies have addressed this problem. In the study by Metz, Isle, Denno, and Odom (1992; see also other references in this study), individuals with these impairments used two experimental home control thermostats with features that allowed easier positioning and viewing. These proved successful, although certain grasping and manipulation strategies used by subjects had not been anticipated by the designers. The appearance of the thermostats also played a role in subject preference.

Another study (Casali, 1992) investigated cursor control device use by physically impaired individuals. The devices were a mouse, trackball, cursor keys, joystick, and tablet. Even those with profound impairments could operate each device with minor device modifications and/or unique operating strategies. The mouse, trackball, and tablet enabled better performance than the keys and joystick. Those with severe impairments performed more slowly than others less disabled, but even the former performed successfully.

Ward (1990) investigated displays in consumer products for the disabled. The data he collected resulted from interviews and home visits. Ward contrasted two different design philosophies: design for the mean (Berns, 1981) and design for all (Kanis, 1988). Following Vanderheiden (1997), our own point of view is that one should (with limitations) design for all because designing for extremes includes design for many people who otherwise fall between the cracks.

8

A Behavioral Theory
of System Design

We can now develop a behavioral theory of how systems are designed. Previously we discussed individual elements of such a theory; now we can integrate them.

The major elements of this theory are: design personnel (the engineering designer and HF specialist), the system to be designed, and the world (e.g., a factory, an external environment) that the system is designed to monitor and/or control. Supporting elements include: the goals set by the customer's wishes, the nature of the design problem that is created by the disparity between those goals and what is presently available to a consumer, the functions to be performed by design personnel (which are primarily cognitive) to solve the design problem, the data available to assist in solving the problem, the tests (both physical and mental) made by the design team in collecting information and to verify hypotheses, the feedback from those tests, and the perceptions of the design personnel that, although mental, are quite real because they have real effects on design processes.

Design begins as a problem to be solved; that is because the customer has specified a technological goal, which immediately becomes a problem because the means of achieving that goal are in most cases not immediately evident to design personnel. There are other goals as well. Designers have the goal of satisfying the technological goal; the specialist has the goal of ensuring that the system will be adequate for whoever operates it. These goals may be initially relatively unclear to design personnel; they contain hidden implications for action that must be extracted from the overall goal and analyzed. These implications may be transformed into subgoals or more immediate goals than the one initially advanced by the customer. Moreover, goals may change during development. Development implies movement over time, progress toward and personnel recognition of design status, and the achievement (or nonachievement) of goals.

System development proceeds in an indeterminate environment. Indeterminism exists when the stimuli that move designers to action may have more than one possible meaning, and the consequences of these various meanings are unclear. As a result, design personnel must engage in a search for information to clarify those meanings. The search also involves a continual testing of the information collected and its meaning relative to goals. This testing is occasionally physical, as in rapid prototyping or usability testing, but is mostly mental as the stimuli are scrutinized for their meaning (meaning turns the raw stimuli into usable information) in relation to the goal. Stimuli that are not goal-relevant are noise and must be discriminated from goal-relevant stimuli.

The theory/model is complicated by the fact that the designer's goal is not the same as the specialist's goal. The goal of the former is functionality as this relates to the system requirement; the goal of the latter is functionality as it relates to the anticipated behavior of the future system operator. These goals must be incorporated as physical mechanisms within the physical system; this means they must be transformed from the symbolic to the physical. Both designers and specialists perform this transformation, but it is more easily performed by the designer because he or she has no need to move design principles and data from one (behavioral) domain to another (physical), as has the specialist. The purpose of design-relevant HF research is to illuminate the transformation process.

Because meanings may not be immediately evident to design personnel, perceptual orientations (ways to look at stimuli) are produced as a result of selecting one or the other of these alternative meanings. The situation is made more difficult because designers are dealing with multiple systems that must be understood. The design project in which personnel work is a system, the equipment being designed is another (real) system, the external environment that the new system will monitor and/or control is another system, and the future operator of the physical system is yet another system. The specialist's responsibility is to incorporate the behavioral qualities of one system (the operator) into another system (the equipment), not literally, but in terms of physical mechanisms that will enable the operator to perform his or her functions effectively and without undue stress. Each system gives rise to a slightly different mental model in the designer, specialist, and operator.

The search for information that we have emphasized results, in part, from the need to understand these different systems and to develop these mental models.

These systems exist in reality, but they are only partially objective because their reality is mediated by the designer's perceptions. Their reality is also affected by the fact that some of these systems (the equipment and operator) only exist in future time. Until a physical artifact is produced at the close of design, the system only exists as an anticipation. Until an actual operator has been trained to operate a physical system, the operator is only the designer's and specialist's concept of the operator.

The physical components of the system have no mentality, but the future operator will have a mentality, which means that he or she will develop and be taught a mental model of the system to perform system operations. The specialist must try to anticipate that model and incorporate it, via physical surrogates, in the physical system. For the operator to be maximally efficient, the operator's mental model should conform as much as possible to the designer's mental model, which the latter builds up during design. The functions performed by personnel during design have both physical and cognitive aspects; for each physical activity listed next, there is a cognitive equivalent that directs and is responsible for the physical one: (a) recognizing that a problem exists; (b) asking questions about the means of accomplishing the design goal; (c) collecting information important to goal accomplishment; (d) analyzing that information and assigning meaning to that information; (e) interrelating (transforming) information within/into the physical system; (f) developing hypotheses about the system being constructed; (g) examining the technological advantages/disadvantages of each hypothesis; (h) testing hypotheses mentally and physically (in continuing evaluations throughout development); (i) recognizing system development status (how far we have progressed toward goal accomplishment); (j) applying criteria of accomplishment (design guidelines, standards) to the products of each design phase; and (k) ceasing the design process (accepting that the technological goal has been achieved).

The preceding is at best only an outline of a behavioral theory of design. A complete description of such a theory would require not only much empirical research, but more than a short text such as this. That is because each design activity is enormously complex. However, why should one expect anything else for a process that is fundamentally responsible for our technological civilization?

Design is further complicated because almost no design involves only one or two individuals. For major systems, hundreds, if not thousands, may be involved, and so one must also consider the organization of the design project. The nature of that organization may help or hinder the effort, and design personnel may lack control over some of its operations.

The entire process is performed in the context of indeterminism and uncertainty, which tends to disorder. The disorder is increased by its reliance on the perception of design personnel, which is inherently imprecise and unstable. On a very abstract level, therefore, one can think of the design process and the functions performed by design personnel as an endeavor to impose order on a disorderly (problematic) world.

We have described design in relatively abstract terms (emphasizing at the same time its concrete manifestations), but it should not be thought that design personnel, in their work, think consciously in terms of the constructs we have developed as explanatory mechanisms. They are too close to the immediate problem.

We have said nothing so far about the necessity to design not only for routine, normal operations, but also for the non-normal, the potential emergencies, the likelihood of a malfunctioning system. Routine operations are by definition more readily predictable because they are designed into the system; emergencies are inherently less predictable because the nature of these emergencies is often only half known. Fail-safe mechanisms to counter all possible failures are almost impossible to include in the system design. The anticipation of emergencies requires design personnel to imagine what might happen (a prediction); the responsibility is even greater for the specialist. The specialist must take the designer's concepts of what and how things might fail and try to imagine how the future operator might respond to the stimuli produced by the impending or actual failure. This requires the specialist to search the operator's mind (one mind imagining another mind), which further complicates the design situation.

If one compares the cognitive elements of this design model with a general model of how humans perform, the similarities (discriminating and interpreting stimuli, making and testing hypotheses, deciding on one, applying criteria, recognizing status, etc.) are quite striking. There is, then, nothing fundamentally unique about design as a behavioral phenomenon. This has the advantage of making it easier to study the design process. There are different emphases, of course: primarily, intense concentration on cognition and the organization of that cognition to solve problems occurring in a context of great uncertainty. However, one could say the same of chess playing or medical diagnosis.

Why, then, should one study design as opposed to chess playing or medical diagnosis? The answer is the purpose of the activity being studied. We hypothesize that design activity determines to a great extent our technological civilization; this makes it more important than chess playing (although only slightly more important than medical diagnosis). Through the study of design, one can perhaps influence design in a positive way and thus modify (only slightly, we fear) our technological civilization.

Beyond this, for the authors, the fascination of design as a process derives in part from its complexity. One wonders how, in view of these complexities, so many systems are developed that, in the end, work fairly well (although somewhat imperfectly). One reason to study the design process is to find out what causes all to end well, if not perfectly. Our theory of design has focused on complexities and problems; it needs to be fleshed out by empirical research so that the critical question of how functioning systems are finally produced can be answered.

All these complexities become research questions. For example, how does complexity in the design process affect the design product? How much complexity in the physical system affects operator performance? How do system designers deal with this complexity in terms of information seeking, hypothesis generation, and testing? Obviously there is much need for research, but other researchers will have other questions to ask.

References

Note. Unless otherwise noted, references cited as *Proceedings* are taken from the *Proceedings* of the annual meetings of the Human Factors and Ergonomics Society.

Adams, J. L. (1984). *Conceptual block busting*. New York: Norton.

Alexander, C. (1979). *The timeless way of building*. New York: Oxford University Press.

Allen, J. J. (1977). *Managing the flow of technology*. Cambridge, MA: MIT Press.

Amabile, T. M. (1983). *The social psychology of creativity*. Berlin: Springer-Verlag.

Amalberti, R. R. (1999). Automation in aviation: A human factors perspective. In D. J. Garland, J. A. Wise, & V. D. Hopkins (Eds.), *Handbook of aviation human factors* (pp. 173–192). Mahwah, NJ: Lawrence Erlbaum Associates.

Ausello, E. F., Lamy, P. P., & Gondek, K. (1990). Generational caregivers, pharmacists and medications: Their contributions to social and personality variables. *Psychology and Aging, 30,* 273–284.

Bainbridge, L. (1981). Mathematical equations or processing routines? In J. Rasmussen & W. B. Rouse (Eds.), *Human detection and diagnosis of system failures* (pp. 259–286). New York: Plenum.

Bainbridge, L. (1991). Mental models in cognitive skill. In A. Rutherford & Y. Rogers (Eds.), *Models in the mind* (chap. 9). New York: Academic Press.

Ballay, J. M. (1987). An experimental view of the design process. In W. B. Rouse & K. R. Boff (Eds.), *System design: Behavioral perspectives on designers, tools, and organizations* (pp. 65–82). New York: North-Holland.

Baron, D., & Levison, W. H. (1977). Display analysis with the optimal control model of the human operator. *Human Factors, 19,* 437–457.

Barr, R. A., & Eberhard, J. W. (1991). Special issue: Safety and mobility of elderly drivers. *Human Factors, 33*(5), 497–600.

Beevis, D., Bost, R., Doring, B., Nordo, E., Papin, J. P., Schuffel, H., & Streets, D. (1992). *Analysis techniques for man-machine design* (Tech. Rep. AC/243, Panel 8, TR7, vols. 1 and 2). Brussels, Belgium: NATO Defense Research Group, Panel 8, RSG14.

Bennett, K. B., Nagy, A. L., & Flach, J. M. (1997). Visual displays. In G. Salvendy (Ed.), *Handbook of human factors and ergonomics* (pp. 659–696). New York: Wiley.

Berliner, D. C., Angell, D., & Shearer, J. W. (1964). Behaviors, measures and instruments for performing evaluation in simulated environments. *Proceedings, Symposium on Quantification of Human Performance.* Albuquerque, NM: University of New Mexico Press.

Berns, T. (1981). The handling of consumer packaging. *Applied Ergonomics, 12,* 153–161.

Bobrow, D. G. (Ed.). (1985). *Qualitative reasoning about physical systems.* Cambridge, MA: MIT Press.

Boff, K. R., & Lincoln, J. E. (1988). *Engineering data compendium: Human perception and performance.* Dayton, OH: Wright-Patterson Air Force Base.

Brunswik, E. (1956). *Perception and the representative design of experiments* (2nd ed.). Berkeley, CA: University of California Press.

Budnick, P. M. (1999). In search of ergonomics knowledge: Effective use of the World Wide Web. In W. Karowski & W. S. Marra (Eds.), *The occupational ergonomics handbook* (pp. 59–66). Boca Raton, FL: CRC Press.

Burns, C. M., & Vicente, K. J. (1996). Judgments about the value and cost of human factors information in design. *Information Processing & Management, 32,* 259–271.

Burns, C. M., Vicente, K. J., Christoffersen, K., & Pawlak, W. S. (1997). Towards viable, useful and usable human factors design guidance. *Applied Ergonomics, 28,* 311–322.

Campbell, J. L. (1998). Evaluation of human factors design guidelines for traveler information systems. *Proceedings,* 1215–1219.

Campbell, J. L., Carney, C., & Kantowitz, B. H. (1999). Developing effective human factors design guidelines: A case study. *Transportation Human Factors, 1*(3), 207–224.

Card, S. K., Moran, T. P., & Newell, A. (1983). *The psychology of human–computer interaction.* Hillside, NJ: Lawrence Erlbaum Associates.

Carroll, J. M. (Ed.). (1995). *Scenario-based design: Envisioning work and technology in system development.* New York: Wiley.

Carroll, J. M., Mack, R. L., & Kellogg, W. A. (1988). Interface metaphors and the user interface design. In M. Helander (Ed.), *Handbook of human–computer interaction* (pp. 322–346). Amsterdam: Elsevier.

Casali, S. P. (1992). Cursor control device use by persons with physical disabilities: Implications for hardware and software design. *Proceedings,* 311–315.

Casali, S. P., & Chase, J. D. (1993). The effects of physical attributes of computer interface design on novice and experienced performance of users with physical disabilities. *Proceedings,* 849–853.

Catani, M. B., & Biers, D. W. (1998). Usability evaluation and prototype fidelity: Users and usability professionals. *Proceedings,* 1331–1335.

Chaffin, D. B. (1997). Human simulation modeling—will it improve ergonomics in design? *Proceedings,* 685–687.

Cody, W. J., Rouse, W. B., & Boff, K. R. (1993). *Automated information management for designers: Functional requirements for computer based associates that support access and use of technical information in design* (Report AL/CF-TR-1993-0069). Dayton, OH: Armstrong Laboratories, Wright-Patterson Air Force Base.

Cooper, A. (1995). *About face: The essentials of user interface design.* Foster City, CA: IDG Books World Wide.

Cooper, G. E., & Harper, R. P., Jr. (1969). *The use of pilot rating in the evaluation of aircraft handling qualities* (Report NASA TN-D-5153). Moffett Field, CA: NASA-Ames Research Center.

Coyne, R. D., Roseman, M. A., Radford, A. D., Balachandran, M., & Gero, J. S. (1990). *Knowledge-based system design.* Reading, MA: Addison-Wesley.

Czaja, S. J. (1990). Special issue: Aging. *Human Factors, 32*(5), 505–620.

Czaja, S. J. (1996). Aging and the acquisition of computer skills. In W. A. Rogers, A. D. Fisk, & N. Walker (Eds.), *Aging and skilled performance: Advances in theory and application* (pp. 201–220). San Diego, CA: Academic Press.

Czaja, S. J. (1997a). System design and evaluation. In G. Salvendy (Ed.), *Handbook of human factors and ergonomics* (pp. 17–40). New York: Wiley.

Czaja, S. J. (1997b). Using technologies to aid the performance of home tasks. In A. D. Fisk & W. A. Rogers (Eds.), *Handbook of human factors and the older adult* (pp. 311–334). San Diego, CA: Academic Press.

Czaja, S. J., Clark, M. C., Weber, R. A., & Nachbar, D. (1990). Computer communication among older adults. *Proceedings, 146*–148.

Davies, J. M. (1998). *A designer's guide to human performance models* (AGARD Advisory Report 356). NATO Research Group paper, Neuilly-sur-Seine, Cedex, France.

Desurive, H. W., Kondziela, J. M., & Attwood, M. E. (1992). What is gained and lost when using evaluation methods other than empirical testing. In G. Monk, D. Diaper, & M. D. Harrison (Eds.), *People and computers VIII* (pp. 89–102). Cambridge, England: Cambridge University Press.

Diaper, D. (1989). *Task analysis for human–computer interaction.* New York: Wiley.

Drury, C. G., Paramore, B., Van Cott, H. P., Grey, S. M., & Corlett, E. N. (1987). Task analysis. In G. Salvendy (Ed.), *Handbook of human factors* (pp. 370–401). New York: Wiley.

Eberhard, J. W., & Barr, R. A. (1992). Special issue: Safety and mobility of elderly drivers, Part II. *Human Factors, 34*(1), 1–120.

Eberts, R. (1997). Cognitive modeling. In G. Salvendy (Ed.), *Handbook of human factors and ergonomics* (pp. 1328–1374). New York: Wiley.

Ellis, R. D., & Cochran, D. L. (1999). Practices to encourage participation of older adults in research and development. *Proceedings, CHI'99,* 39–40.

Endsley, M. (1995). Towards a theory of situation awareness. *Human Factors, 37,* 32–64.

Fisk, A. D. (1999). Human factors and the older adult. *Ergonomics in Design, 7*(1), 8–13.

Fitts, P. M. (Ed.). (1951). *Human engineering for an effective air-navigation and traffic-control system.* Columbus, OH: Ohio State University Research Foundation.

Flach, J. M., Vicente, K. M., Tanabe, F., Monta, K., & Rasmussen, J. (1998). An ecological approach to interface design. *Proceedings,* 295–299.

Fleishman, E. A., & Quaintance, M. K. (1984). *Taxonomies of human performance: The description of human tasks.* Orlando, FL: Academic Press.

Flower, S. (1996). *Software failure: Management failure: Amazing stories and cautionary tales.* New York: Wiley.

Fozzard, J. L. (Ed.). (1981). Special issue: Aging. *Human Factors, 23*(1).

Fu, L., Salvendy, G., & Turley, L. (1998). Who finds what in usability evaluation. *Proceedings,* 1341–1345.

Geer, C. W. (1981). *Human engineering procedures guide* (Report AFAMRL-TR-81-35). Wright-Patterson Air Force Base: Aerospace Medical Division.

General Accounting Office. (1981). *Effectiveness of US forces can be increased through improved weapon system design* (Report PSAD-81-17). Washington, DC: Author.

Gibson, J. J. (1979). *The ecological approach to visual perception.* Boston, MA: Houghton-Mifflin.

Gillian, D. J., & Cooke, N. J. (1998). Making usability data more usable. *Proceedings,* 300–304.

Goodstein, L. P., Andersen, H. B., & Olsen, S. E. (Eds.). (1988). *Tasks, errors and mental models.* London: Taylor & Francis.

Gruber, T. R., & Russell, D. M. (1996). Generative design rationale beyond the record and replay paradigm. In T. P. Moran & J. M. Carroll (Eds.), *Design rationale: Concepts, techniques, and use* (pp. 323–350). Mahwah, NJ: Lawrence Erlbaum Associates.

Hale, C. R., & Rowe, A. L. (1998). Developing design requirements from cognitive task data: Replacing magic with methodology. *Proceedings,* 291–293.

Harris, J., & Henderson, A. (1999). A better mythology for system design. *Proceedings, CHI'99*, 88–95.

Hartson, H. R., & Hix, D. (1989). Human–computer interface development: Concepts and systems for its management. *ACM Computing Surveys, 21*(1), 5–93.

Hayak, U. S. L. (1995, January). Elders-led design. *Ergonomics in Design*, pp. 8–13.

Helander, M. (1988). *Handbook of human-computer interaction*. Amsterdam: Elsevier.

Hendrick, H. W. (1984). Cognitive complexity, conceptual systems, and organizational design and management: Review and ergonomic implications. In H. W. Hendrick & O. Brown (Eds.), *Human factors in organizational design and management* (pp. 467–478). Amsterdam: North-Holland.

Hendrick, H. W. (1996). All member survey: Preliminary results. *HFES Bulletin, 36*(6), 1, 4–6.

Hendrick, H. W. (1997). Organizational design and macroergonomics. In G. Salvendy (Ed.), *Handbook of human factors and ergonomics* (pp. 594–636). New York: Wiley.

Hoffman, R. R., & Woods, D. D. (2000). Studying cognitive systems in context: Preface to the special section. *Human Factors, 42*, 1–7.

Hofmeester, K., Dunne, A., Carver, B., Susani, M., & Pacenti, E. (1999). A modern rule for the village elders. *Proceedings, CHI'99*, 43–44.

Hollis, W. W. (1998). Simulation based acquisition—some thoughts on things not to forget. *Phalanx, 31*(4), 28–29.

Hudspith, S. (1997). Beyond utility: A framework for designing user experience. *Proceedings*, 447–450.

Isensee, S., Roberts, D., Berry, D., & Mullaly, J. (1998). Object-oriented user interface design with OVID. *Proceedings*, 310–314.

Jamieson, B. A., & Rogers, W. A. (1998). Age-related differences in the acquistion, transfer and retention of using an automatic teller machine. *Proceedings*, 156–160.

Jeffries, R., Miller, J. R., Wharton, C., & Uyeda, K. M. (1991). User-interface evaluation in the real world: A comparison of four techniques. *Proceedings, ACM CHI'91*, 119–124.

John, B. E. (1988). *Contributions to engineering models of human–computer interaction*. Unpublished PhD dissertation, Carnegie Mellon University, Pittsburgh, PA.

Jones, J. C. (1980). *Design methods: Seeds of human futures*. New York: Wiley.

Kanis, H. (1988). Design for all? The use of consumer products by the disabled. *Proceedings*, 416–419.

Keen, P. G. W., & Scott Morton, M. S. (1978). *Decision support systems: An organizational perspective*. Reading, MA: Addison-Wesley.

Kieras, D. E. (1988). Towards a practical GOMS model methodology for user interface designs. In M. Helander (Ed.), *Handbook of human–computer interaction* (pp. 135–157). Amsterdam: Elsevier.

Kieras, D. E., & Polson, P. G. (1985). An approach to the formal analysis of user complexity. *International Journal of Man–Machine Studies, 32*, 365–394.

Kira, A. (1976). *The bathroom*. New York: Viking.

Kirwan, B., & Ainsworth, L. K. (1992). *A guide to task analysis*. London: Taylor & Francis.

Klein, G. A. (1993). A recognition-primed decision (RPD) model of rapid decision making. In G. A. Klein, J. Orasanu, R. Calderwood, & C. E. Zambo (Eds.), *Decision making in action: Models and methods* (pp. 138–147). Norwood, NJ: Ablex.

Koffka, K. (1935). *Principles of Gestalt psychology*. New York: Harcourt & Brace.

Kramer, A. F., Larish, J. F., & Strayer, D. L. (1995). Training for attentional control of individual task settings: A comparison of young and old adults. *Journal of Experimental Psychology: Applied, 1*, 50–76.

Kurke, M. I. (1961). Operational sequence diagrams in system design. *Human Factors, 3*(1), 66–78.

Laughery, K. (1989). Micro-SAINT: A tool for modeling human performance in systems. In G. McMillan et al. (Eds.), *Applications of human performance models to system design* (pp. 219–230). New York: Plenum.

Laughery, K. R., Jr., & Corker, K. (1997). Computer modeling and simulation. In G. Salvendy (Ed.), *Handbook of human factors and ergonomics* (pp. 1378–1408). New York: Wiley.

Laughery, K. R., Sr., & Laughery, K. R., Jr. (1987). Analytic techniques for function analysis. In G. Salvendy (Ed.), *Handbook of human factors* (pp. 329–354). New York: Wiley.

Lee, S. (1995). Access to computer systems with graphical user interface by touch: Haptic discrimination of icons. *Proceedings,* 742–746.

Lee, J. D., & Moray, N. (1992). Operators' monitoring patterns and fault recovery in the supervisory control of semi-automatic process. *Proceedings,* 1143–1147.

Leifer, L. (1987). On nature of design and an environment for design. In W. B. Rouse & K. R. Boff (Eds.), *System design: Behavioral perspectives on designers, tools, and organizations* (pp. 211–220). New York: North-Holland.

Leirer, V. O., Morrow, T. G., Tanke, E. D., & Panante, G. M. (1991). Elders non-adherence: Its assessment and medication reminding by voice mail. *The Gerontologist, 31*(4), 514–520.

Lind, M. (1988). System concepts and the design of man–machine interfaces for supervisory control. In L. P. Goodstein, H. B. Andersen, & S. E. Olsen (Eds.), *Tasks, errors and mental models* (pp. 269–277). London: Taylor & Francis.

Liu, Y. (1994). A queuing network model of human performance of concurrent spatial and verbal tasks. *Proceedings, 1994 International Conference on Systems, Man and Cybernetics* (pp. 2761–2766). New York: Institute of Electrical and Electronics Engineers.

Liu, Y. (1996). Queuing network modeling of elementary mental processes. *Psychological Review, 103,* 116–136.

Liu, Y. (1997). Software-user interface design. In G. Salvendy (Ed.), *Handbook of human factors and ergonomics* (pp. 1689–1724). New York: Wiley.

Luczak, H. (1997). Task analysis. In G. Salvendy (Ed.), *Handbook of human factors and ergonomics* (pp. 340–416). New York: Wiley

Lund, A. M. (1997). Expert ratings of usability maxims. *Ergonomics in Design, 5*(3), 15–20.

MacLean, A., Young, R. M., Bellotti, V., & Moran, T. P. (1996). Elements of design space analysis. In T. P. Moran & J. M. Carroll (Eds.), *Design rationale: Concepts, techniques and use* (pp. 73–102). Mahwah, NJ: Lawrence Erlbaum Associates.

Maltz, M., & Shinar, D. (1999). Eye movements of younger and older drivers. *Human Factors, 41*(6), 15–25.

Mannova, B. (1999). Teaching IT for seniors. *Proceedings, CHI'99,* 49–50.

Maslow, A. H. (1954). *Motivation and personality.* New York: Harper.

Matthews, M. L., & McFadden, S. M. (1993). Implications of the user's information processing strategy on the design of decision aids for complex systems. *Proceedings,* 358–362.

McCracken, J. R. (1990). *Questions: Assessing the structure of knowledge and the use of information in design problem solving.* Unpublished PhD dissertation, Ohio State University, Columbus, OH.

McCrary, F. A., & Williges, R. C. (1998). Effects of age and field of view on spatial learning in an immersive virtual environment. *Proceedings,* 1491–1495.

McKim, R. H. (1980). *Visual thinking.* Lifetime Learning Publications (cited in Leifer).

McNeese, M. D. (1998). New frontiers in cognitive task analysis: Bridging the gap between cognitive analysis and cognitive engineering. *Proceedings,* 377–379.

Mead, S. E., & Fisk, A. D. (1997). Effects of matching cognitive and perceptual motor training to task components in complex task performance by older and younger adults. *Proceedings,* 115–119.

Mead, S. E., & Fisk, A. D. (1998). Measuring skill acquisition and retention with an ATM simulator: The need for age-specific training. *Human Factors, 40,* 516–523.

Meister, D. (1971). *Human factors: Theory and practice.* New York: Wiley.

Meister, D. (1985a). *Behavioral analysis and measurement methods.* New York: Wiley.

Meister, D. (1985b). The two worlds of human factors. In R. E. Eberts & C. G. Eberts (Eds.), *Trends in human factors/ergonomics II* (pp. 3–11). Amsterdam: Elsevier.

Meister, D. (1987). A cognitive theory of design and requirements for a behavioral design aid. In W. B. Rouse & K. R. Boff (Eds.), *System design: Behavioral perspectives on designers, tools and organizations* (pp. 211–220). New York: North-Holland.

Meister, D. (1991). *The psychology of system design.* Amsterdam: Elsevier.

Meister, D. (1996a). Cognitive behavior of nuclear reactor operators. *International J. Industrial Ergonomics, 16,* 109–122.

Meister, D. (1996b). A new theoretical structure for developmental ergonomics. *Proceedings, 4th Pan Pacific Conference on Occupational Ergonomics* (pp. 688–690). Taipei, Taiwan.

Meister, D. (1997). Studies in the history of human factors ergonomics. *Proceedings,* 548–550.

Meister, D. (1999). *The history of human factors and ergonomics.* Mahwah, NJ: Lawrence Erlbaum Associates.

Meister, D., & Farr, D. E. (1967). The utilization of human factors information by designers. *Human Factors, 9,* 71–87.

Metz, S., Isle, B., Denno, S., & Odom, J. (1992). Thermostats for individuals with movement disabilities: Design options and manipulation strategies. *Proceedings,* 180–184.

Meyer, B., & Fisk, A. D. (1998). Toward an understanding of age-related use of incidental consistency. *Proceedings,* 161–165.

Miller, C. A., & Vicente, K. J. (1998). Toward an integration of task and work-domain analysis techniques for human–computer interface design. *Proceedings,* 336–340.

Miller, G. (1956). The magical number 7 plus or minus two: Some limits on our capacity for processing information. *Psychological Review, 63,* 81–97.

Miller, R. B. (1953). *A method for man–machine analysis* (Report 53-137). Wright-Patterson Air Force Base, OH: Wright Air Research and Development Command.

MIL-STD 1472F. (1999). *Human engineering design criteria for military systems, equipment, and facilities.* Redstone Arsenal, AL: U.S. Army Missile Command.

Monk, D. L., Swierenga, S. J., & Lincoln, J. E. (1992). Developing behavioral phenomena test benches. *Proceedings,* 1106–1109.

Moran, T. P. (1981). The command language grammar. *International Journal of Man–Machine Studies, 15,* 3–50.

Moran, T P., & Carroll, J. M. (Eds.). (1996). *Design rationale: Concepts, techniques, and use.* Mahwah, NJ: Lawrence Erlbaum Associates.

Moray, N. (1986). Monitoring behavior and supervisory control. In K. R. Boff (Ed.), *Handbook of perception and human performance* (pp. 40/1–40/51). New York: Wiley.

Moray, N. (1994). "De maximus non curat lex" or how context reduces science to art in the practice of human factors. *Proceedings,* 526–530.

Moray, N. (1997). Human factors in process control. In G. Savendy (Ed.), *Handbook of human factors and ergonomics* (pp. 1944–1971). New York: Wiley.

Morphew, M. E., Thordsen, M. L., & Klein, G. (1998). The development of cognitive task analysis methods to aid interface design. *Proceedings,* 305–309.

Mostow, J. (1985, Spring). Towards better models of the design process. *The AI Magazine,* pp. 44–57.

Munger, S., Smith, R. W., & Payne, D. (1962). An index of electronic equipment operability: Data store (Report AIR-C43-1/62-RP[1]). Pittsburgh, PA: American Institute for Research.

Murch, G. M. (1987). Color graphics: Blessing or ballyhoo? In R. M. Baecker & W. A. S. Buxton (Eds.), *Readings in human–computer interaction: A multidisciplinary approach* (pp. 333–341). San Mateo, CA: Morgan Kauffman.

Murphy, R. A., & Basili, A. (1993). Developing the user-system interface for a communications system for ALS patients and others with severe neurological impairments. *Proceedings*, 854–858.

Murray, J. L., & Liu, Y. (1995a). *Towards a distributed intelligent agent architecture for human–machine systems in hortatory operations* (Tech. Rep. 95-17). Ann Arbor: University of Michigan.

Murray, J. L., & Liu, Y. (1995b). *The colloquium: Ontological support for hortatory operations* (Tech. Rep. 95-18). Ann Arbor: University of Michigan.

Mynatt, E. D., Adler, A., Ito, M., Linde, C., & O'Day, V. L. (1999). Learning from seniors in network communities. *Proceedings, CHI'99*, 47–48.

Newman, R. A. (1999). Issues in defining human roles and interactions in systems. *Systems Engineering, 2*(3).

Nielsen, J. (1990). Paper versus computer implementations as mockup scenarios for heuristic evaluation. *Proceedings, IFIP INTERACT'90: Human Computer Interaction*, 315–320.

Nielsen, J. (1994). Usability laboratories. *Behavioral and Information Technology, 13*, 3–8.

Nielsen, J. (1997). Usability testing. In G. Salvendy (Ed.), *Handbook of human factors and ergonomics* (pp. 1543–1568). New York: Wiley.

Nielsen, J., & Molich, R. (1990). Heuristic evaluation of user interfaces. *Proceedings, ACM CHI'90 Conference*, 249–256.

Norman, D. A. (1983). Some observations on mental models. In D. Gentner & A. I. Stevens (Eds.), *Mental models* (pp. 78–92). Hillsdale, NJ: Lawrence Erlbaum Associates.

Norman, D. A. (1986). Cognitive engineering. In D. A. Norman & S. W. Draper (Eds.), *User centered system design: New perspectives on human–computer interaction*. Hillsdale, NJ: Lawrence Erlbaum Associates.

Norman, D. A., & Draper, S. W. (Eds.). (1986). *User centered system design: New perspectives on human-computer interaction*. Hillsdale, NJ: Lawrence Erlbaum Associates.

O'Brien, T. G., & Charlton, S. G. (1996). *Handbook of human factors testing and evaluation*. Mahwah, NJ: Lawrence Erlbaum Associates.

Olson, G. M., Olson, J. S., Storrosten, M., Carter, M., Hertsleb, J., & Reuter, H. (1996). The structure of activity during design meetings. In T. P. Moran & J. M. Carroll (Eds.), *Design rationale: Concepts, techniques and use* (pp. 217–235). Mahwah, NJ: Lawrence Erlbaum Associates.

Osborn, A. (1963). *Applied imagination*. New York: Scribner & Sons.

Pasmore, W. A. (1988). *Designing effective organizations: The socio-technical system perspective*. New York: Wiley.

Pattipatti, K. R., & Kleinman, D. L. (1992). A review of the engineering models of information processing and decision making in multi-task supervisory control. In D. Damos (Ed.), *Multiple task performance* (pp. 35–68). London: Taylor & Francis.

Payne, S. J., & Green, T. R. G. (1986). Task-action grammars: A model of the mental representation of task languages. *Human–Computer Interaction, 2*(2), 93–133.

Pejtersen, A. M. (1985). Implications of users' value perception for the design of a bibliographical retrieval system. In J. C. Agrawal & P. Zunde (Eds.), *Empirical foundations of information and software sciences* (pp. 23–37). New York: Plenum.

Pew, R. W., & Mavor, A. S. (Eds.). (1998). *Modeling human and organizational behavior: Application to military simulations*. Washington, DC: National Academy Press.

Pirkl, J. J. (1994). *Transgenerational design: Products for an aging population*. New York: Van Nostrand Reinhold.

Potter, S. S., Roth, E. M., Woods, D. D., & Elm, W. C. (1998). A framework for integrating cognitive task analysis into the system development process. *Proceedings*, 395–399.

Potts, C. (1996). Supporting software design: Integrating design methods and design rationale. In T. P. Moran & J. M. Carroll (Eds.), *Design rationale: Concepts, techniques and use* (pp. 295–322). Mahwah, NJ: Lawrence Erlbaum Associates.

Rasmussen, J. (1983). Skills, rules, and knowledge: Signals, signs and symbols and other distinctions in human performance models. *IEEE Transactions on Systems, Man, and Cybernetics, SMC-132,* 257–266.

Rasmussen, J. (1986). *Information processing and human-machine interaction: An approach to cognitive engineering.* New York: North-Holland.

Rasmussen, J., Pejtersen, A. M., & Goodstein, L. P. (1994). *Cognitive system engineering.* New York: Wiley.

Rasmussen, J., & Vicente, K. J. (1989). Coping with human error through system design: Implications for ecological interface design. *IEEE Transactions on Man-Machine Studies, 31,* 517–534.

Reisner, P. (1981). Formal grammar and human factors design of an interactive system. *IEEE Transactions on Software Engineering, SE-7*(2), 229–240.

Robertson, G. L., & Hix, D. (1994). User interface design guidelines for computer accessibility by mentally retarded adults. *Proceedings,* 300–304.

Rogers, W. A. (1999). Technology training for older adults. *Proceedings, CHI'99,* 51–52.

Rogers, W. A., & Fisk, A. D. (1997). Automatic teller machines: Design and training issues. *Ergonomics in Design, 5*(1), 4–9.

Rogers, W. A., Fisk, A. D., Mead, S. E., Walker, N., & Cabrera, E. F. (1996). Training older adults to use automatic teller machines. *Human Factors, 38,* 425–433.

Rogers, W. A., Meyer, B., Walker, N., & Fisk, A. D. (1998). Functional limitations to daily living tasks in the aged: A focus group analysis. *Human Factors, 40,* 111–125.

Rouse, W. B. (1980). *Systems engineering models of human–machine interaction.* Amsterdam: Elsevier.

Rouse, W. B. (1986). On the value of information in system design: A framework for understanding and aiding designers. *Information Processing & Management, 22,* 279–285.

Rouse, W. B., & Boff, K. R. (Eds.). (1987). *System design: Behavioral perspectives on designers, tools, and organizations.* New York: North-Holland.

Rouse, W. B., Cody, W. R., & Boff, K. R. (1991). The human factors of system design: Understanding and enhancing the role of human factors engineering. *International Journal of Human Factors in Manufacturing, 1,* 87–104.

Rouse, W. B., Cody, W. R., Boff, K. R., & Frey, P. R. (1990). Information systems for supporting design of complex human machines. In C. T. Leondes (Ed.), *Advances in aeronautical systems* (pp. 41–100). Orlando, FL: Academic Press.

Rubinstein, R., & Hersh, H. (1984). *The human factor: Designing computer systems for people.* Maynard, MA: Digital Press.

Sage, A. P. (1987). Knowledge, skills and information requirements for system design. In W. B. Rouse & K. R. Boff (Eds.), *System design: Perspectives on designers, tools and organizations* (pp. 285–304). New York: North-Holland.

Salthouse, T. A., Fristoe, N. M., Lineweaver, T. T., & Coon, V. E. (1995). Aging of attention: Does the ability to divide decline? *Memory and Cognition, 23,* 59–71.

Salvendy, G. (Ed.). (1997). *Handbook of human factors and ergonomics.* New York: Wiley.

Scerbo, M. W. (1995). Usability testing. In J. Weimer (Ed.), *Research techniques in human engineering* (pp. 72–111). Englewood Cliffs, NJ: Prentice-Hall.

Schroder, H. M., Driver, M. J., & Streufert, S. (1967). *Human information processing.* New York: Holt, Rinehart & Winston.

Schvaneveldt, R. W., Durso, F. T., & Dearholt, D. W. (1985). *Pathfinder: Scaling with network structures* (CRL Memoranda Series, Report MCCS-5-85-9). Los Cruces, NM: New Mexico State University Press.

Schweikert, R. (1978). A critical path generalization of the additive factor method: Analysis of a Stroop task. *Journal of Mathematical Psychology, 18,* 105–139.

Sheridan, T. B. (1984). Supervisory control of remote manipulators. In W. B. Rouse (Ed.), *Advances in man–machine system research* (Vol. I, pp. 49–137). New York: JAI.

Sheridan, T. B. (1998). Allocating functions rationally between humans and machines. *Ergonomics in Design, 6*(3), 20–25.

Shneiderman, B. (1983). Direct manipulation: A step beyond programming languages. *IEEE Computer, 16*, 57–69.

Shneiderman, B. (1998). *Designing the user interface: Strategies for effective human-computer interaction* (3rd ed.). Also, 2nd edition, 1992. Reading, MA: Addison-Wesley.

Siegel, A. I., & Wolf, J. J. (1969). *Man–machine simulation models*. New York: Wiley.

Singley, M. K., & Carroll, J. M. (1996). Synthesis by analysis: Five modes of reasoning that guide design. In T. P. Moran & J. M. Carroll (Eds.), *Design rationale: Concepts, techniques, and use* (pp. 241–266). Mahwah, NJ: Lawrence Erlbaum Associates.

Smith, J. M. (1987). Intuition by design. In W. B. Rouse & K. R. Boff (Eds.), *System design: Perspectives on systems, tools and organizations* (pp. 305–318). New York: North-Holland.

Smith, P. J., McCoy, E., & Layton, C. (1993). Design-induced error in flight planning. *Proceedings*, 1091–1095.

Smith, S. L. (1988). Standards versus guidelines for designing user interface software. In M. Helander (Ed.), *Handbook of human–computer interaction* (pp. 877–889). Amsterdam: Elsevier.

Smith, S. L., & Mosier, J. W. (1986). *Guidelines for designing user interface software* (Tech. Rep. MTR-10090, ESD-TR-86-278). Bedford, MA: MITRE Corporation.

Stanney, K. M., Maxey, T. L., & Salvendy, G. (1997). Socially centered design. In G. Salvendy (Ed.), *Handbook of human factors and ergonomics* (pp. 637–656). New York: Wiley.

Staplin, L. K., & Fisk, A. D. (1991). Left-turn intersection problems: A cognitive engineering approach to improve the safety of young and old drivers. *Human Factors, 33*, 559–571.

Sweetland, J. H. (1988). Beta tests and end-user surveys: Are they valid? *Database, 11*(1), 27–37.

Taylor, F. W. (1919). *Principles of scientific management*. New York: Harper.

Tullis, T. S. (1988). Screen design. In M. Helander (Ed.), *Handbook of human–computer interaction* (pp. 377–411). Amsterdam: Elsevier.

Vanderheiden, G. C. (1997). Design for people with functional limitations resulting from disability, aging, or circumstance. In G. Salvendy (Ed.), *Handbook of human factors and ergonomics* (pp. 2010–2052). New York: Wiley.

Van Gigch, J. F. (1974). *Applied general system theory*. New York: Harper & Row.

VanGundy, A. B. (1981). *Technique of structured problem solving*. New York: Van Nostrand Reinhold.

Vicente, K. J. (1999). *Cognitive work analysis: Towards safe, productive, and healthy computer-based work*. Mahwah, NJ: Lawrence Erlbaum Associates.

Vicente, K. J., Burns, C. M., & Pawlak, W. S. (1993). *Egg-sucking, mousetraps, and the Tower of Babel: Making human factors guidance more accessible to designers* (Report CEL 93-01). Toronto, Canada: University of Toronto, Cognitive Engineering Laboratory.

Vicente, K. J., & Rasmussen, J. (1992). Ecological interface design: Theoretical foundations. *IEEE Transactions on Systems, Man, and Cybernetics, SMC-22*, 589–606.

Virzi, R. A. (1989). What you can learn from a low fidelity prototype. *Proceedings*, 224–228.

Walker, N., Millians, J., & Worden, A. (1996). Mouse accelerations and performance of older computer-users. *Proceedings*, 151–154.

Walker, N., Philbin, D. A., & Fisk, A. D. (1997). Age-related differences in movement control: Adjusting submovement structure to optimize performance. *Journal of Gerontology: Psychological Sciences, 52B*, 40–52.

Ward, J. T. (1990). Designing consumer product displays for the disabled. *Proceedings*, 448–451.

Wasserman, A. I. (1985). Extending state transmission diagrams for the specification of human–computer interaction. *IEEE Transactions on Software Engineering, 11*, 699–713.

Weimer, J. (Ed.). (1995). *Research techniques in human engineering*. Englewood Cliffs, NJ: Prentice-Hall.

Whitaker, L. A., & Moroney, W. F. (1992). Improving the interface between human factors data and designers: Exemplified in a CASHE reaction time prototype. *Proceedings*, 1092–1095.

Whitefield, A., Wilson, F., & Dowell, J. (1991). A framework for human factors evaluation. *Behavior & Information Technology, 10*(1), 65–79.

Wickens, C. D. (1992). *Engineering psychology and human performance* (2nd ed.). New York: HarperCollins.

Wickens, C. D., & Carswell, M. (1995). The proximity compatibility principle: Its psychological foundation and relevance to display design. *Human Factors, 37*(3), 473–494.

Wiener, E. L., & Nagel, D. C. (Eds.). (1988). *Human factors in aviation*. San Diego, CA: Academic Press.

Williges, B. H., & Williges, R. C. (1984). Dialogue design considerations for interactive computer systems. In F. A. Muckler (Ed.), *Human factors review-1984* (pp. 167–208). Santa Monica, CA: The Human Factors Society.

Williges, R. C., Williges, B. H., & Elkerton, J. (1987). Software interface design. In G. Salvendy (Ed.), *Handbook of human factors* (pp. 1416–1449). New York: Wiley.

Wilson, J., & Rosenberg, D. (1988). Rapid prototyping for user interface design. In M. Helander (Ed.), *Handbook of human–computer interaction* (pp. 859–875). Amsterdam: Elsevier.

Wise, B. K., & Wise, J. A. (1991). Children's human factors in the design of a preschool educational furnishings system. *Proceedings*, 541–545.

Woods, D. D., & Roth, E. M. (1988). Cognitive systems engineering. In M. Helander (Ed.), *Handbook of human–computer interaction* (pp. 3–43). New York: Elsevier.

Woods, D. D., Watts, J. C., Graham, J. M., Kidwell, D. L., & Smith, P. J. (1996). Teaching cognitive systems engineering. *Proceedings*, 259–263.

Woodson, W. (1954). *Human engineering guide for equipment designers*. Berkeley, CA: University of California Press.

Woodson, W., Tillman, B., & Tillman, P. (1992). *Human factors design handbook: Information and guidelines for the design of systems, facilities, equipment, and products for human use* (2nd ed.). New York: McGraw-Hill.

Worden, A., Walker, N., Bharat, K., & Hudson, S. (1997). Making computers easier for older adults to use: Area cursors and sticky icons. *Proceedings*, Human Factors in Computing Systems '97 meeting. New York: ACM.

Zipf, G. K. (1965). *Human behavior and the principle of least effort* (2nd ed.). New York: Hafner.

Author Index

Subject Index

For Product Safety Concerns and Information please contact our EU representative GPSR@taylorandfrancis.com Taylor & Francis Verlag GmbH, Kaufingerstraße 24, 80331 München, Germany